T0176684

AUTOMATED TRANSIT

AUTOMATED TRANSIT
Planning, Operation, and Applications

RONGFANG (RACHEL) LIU

IEEE PRESS

WILEY

For general information on our other products and services or for technical support, please contact our Customer Care Department within the United States at (800) 762-2974, outside the United States at (317) 572-3993 or fax (317) 572-4002.

Wiley also publishes its books in a variety of electronic formats. Some content that appears in print may not be available in electronic formats. For more information about Wiley products, visit our web site at www.wiley.com.

Library of Congress Cataloging-in-Publication Data is available.

ISBN: 978-1-118-89100-1

Printed in the United States of America

10 9 8 7 6 5 4 3 2 1

This book is dedicated to the three men in my life:

ZHONG:
My rock, who supports anything I am willing to explore;

LYNDALL:
My conscious, who shows me that there might be
another side to any story;

CHARLIE:
My lucky star, who makes me believe that
there are always roads under my feet …

CONTENTS

FOREWORD

The science of automated transit is relatively young. Although people have explored travel options since the early days of history, it is only in the last 50 years or so that engineers and scientists have unveiled transportation options that are fully automated. From driverless autos to personal rapid transit designs to full-functioning extended people mover systems, we are learning to give up the driver's seat and trust the power of smart technology.

When we began Lea+Elliott in the 1970s, specializing in automated people movers was an anomaly. Some engineers could not understand why we would focus on such a niche market. At that time, the industry was mostly focused on transporting passengers quickly, safely, and efficiently between terminals in large airports. Today, as we work on nearly every people mover system in the world, we know that the early technology provided the impetus for systems that are literally changing how we think about travel. For example, consider Honolulu, HI. Today, the City and County of Honolulu, in cooperation with the Federal Transit Administration (FTA), is implementing a 20-mile-long automated metro rail system that will serve 21 passenger stations. It will be the first automated metro light rail system in the United States since JFK AirTrain and will truly change lives for people within its reach.

In such a rapidly changing transit environment, Dr. Rongfang (Rachel) Liu is the logical person to create this book on the state of automated transit— and to show where it will lead us in the days to come. As a professional engineer, licensed planner, and professor in the Department of Civil and Environmental Engineering at New Jersey Institute of Technology, Dr. Liu brings so much more to the transit discussion. Her vast research has been published in many books, book chapters, and transportation journals. Her additional skills in modeling and expertise in intermodal research further round out her understanding of this complex and multi-faceted transportation technology. I trust that her knowledge and perspective, provided in these pages, will offer insights to help you better understand the sophisticated systems that make automated transit so fascinating. Her thoughts may well spur your thoughts which just might take automated transit technology to the next level. Enjoy!

JACK NORTON

President/CEO
Lea+Elliott, Inc.
Dallas/Fort Worth, TX

PREFACE

The idea for this book was conceived a few years ago when I wrote a book chapter titled "The Spectrum of Automated Guideway Transit and Its Applications," which is published in the *Handbook of Transportation Engineering* (Kutz, 2011). I have accumulated a large amount of information and felt that there are so much more can be said but has not been included in the chapter due to the length limit. In fact, so much was misunderstood or misconstrued for automated guideway transit (AGT) all together for the past half of a century.

As the Committee Chair for AP040: Automated Transit Systems (ATS), Transportation Research Board (TRB), I have been working with my committee members and an array of stakeholders, which include transit and Airport Automated People Mover (AAPM) operators, local and federal government agencies, and private entities such as Google, CarShare Inc., and Ride Scott. Continuing dialogs among the automated transit community made the critical needs paramount. I felt strongly that it is time to have a thorough examination of the automated transit technology development and its applications.

Furthermore, the promises and expectations created by the "future transportation" starting in 1970s need to be evaluated after more than four decades. The successful implementations of automated transit in various international locations, such as Paris, Toronto, London, and Kuala Lumpur, and the apparent lack of automated transit applications in the urban environment in the United States warrant in-depth analyses. The ultimate lessons learnt via various not so successful concepts, ideas, and programs are also valuable for an emerging new paradigm, such as automated transit or driverless vehicles, to grow and prosper.

The rapid development of driverless cars by Google and others not only grabbed the attention of the US Congress, which held a hearing on "the future role of autonomous vehicles in US transportation" in October 2013, but also created a perfect opportunity to have a thorough examination of automated transit applications and their impact and implications for our society. As pointed out by an anonymous proposal reviewer, "It is the right time to have a book on automated transit system (ATS). There are many different automated transit systems worldwide. A book on this topic will be of interest to transportation professionals, researchers, and some graduate students to learn basic concepts, technologies and successful examples related to ATS."

The basic structure of this book follows typical technology development document that begins with a brief definition of the automated transit, Chapter 1, and their historical development in Chapter 2. After a thorough description of the technical specifications in Chapter 3, the manuscript highlights a few representative applications from each sub-group of automated transit spectrum in Chapter 4. The case studies around the world not only showcase different technologies and their applications but also identify the vital factors that affect each system and performance evaluations of existing applications in Chapters 5 and 6. Chapters 7 and 8 of the the book is devoted to planning and operation of automated transit applications in both macro and micro levels. The last two chapters of the book highlight the lessons learnt from the past experiences and try to project the new paradigm shift from the current, conventional transportation systems.

ACKNOWLEDGMENTS

My sincere appreciation goes to many people who contributed directly and indirectly to the fruition of this book. First, I am indebted to many members and friends of Automated Transit Systems (ATS – AP040) Committee, previously Major Activity Center Circulation Systems, Transportation Research Board. During my 6-year term as the committee chair starting in 2008, some in-depth discussions and dialogs on the directions and structure of the committee have propelled me to explore the historical development of automated transit technology and engage wide ranges of stakeholders, which provided a rich background to shape my general view of automated transit development.

Second, I would like to express my gratitude to the visionary leaders who organized the Automated Vehicle Symposium (AVS), especially those who contributed to the Automated Transit and Shared Mobility Track (ATSM), which are co-sponsored by the same TRB ATS committee. The exponential growth in AVS attendees during the past 5 years and pointed discussions and/or arguments have instigated further research, clarification, and selection of book contents.

Next, my appreciation goes to the few selected colleagues and friends who have painstakingly reviewed the manuscript and provided valuable suggestions and improvements: Stanley E. Young, University of Maryland; William J. Sproule, Michigan Technological University; Gary Hsue, Arup, Inc.; Ingmar Andreasson, Royal Institute of Technology, Sweden; Wayne Cottrell, California State University at Pomona; Sam Lott, Kimbley Horn and Associates, Inc.; Naderah Moini, University of Illinois at Chicago; Jerry Lutin, New Jersey Transit, retired; Larry Fabian, Trans.21, Inc.; Peter Muller, PRT Consultant,

LLC; Alex Lu, New York City Metropolitan Transportation Authority; Walter Kulyk, Federal Transit Administration, retired; Ruben Juster, University of Maryland; and Matthew Lash, Noblis Inc.

Last but not least, I wish to thank my colleagues in New Jersey Institute of Technology (NJIT), who granted me a 1-year paid sabbatical leave. The sabbatical leave not only shielded me from regular teaching load, daily commute, and trivia administrative duties but also liberated my mind and spirit to think deeply, reach widely, and explore freely. My sincere appreciation goes to my former students, Zhaodong (Tony) Huang, now with Ningbo University, and Hongmei Cao, now with Inner Mongolia University, who helped a great deal in compiling graphs, tables, and glossaries in conjunction with tedious document review and edit.

I have accumulated a large number of photographs, tables, and figures via previous research and project experiences and have tried to provide appropriate credit to the maximum extent possible. I regret any errors or oversights in crediting any material, if any. Of course, any other errors, omissions, and oversights are my responsibility and will be corrected once known.

RONGFANG (RACHEL) LIU

ABBREVIATIONS

AA   alternative analysis (7)
AADT   annual average daily traffic (6.3)
AAPM   airport automated people mover (2.1)
AB   automated bus (1.2)
AC   alternating current (2.4.1)
ACRP   Airport Cooperative Research Program (10.5)
AG   automated guideway (6.1)
AGRT   advanced group rapid transit (2.2)
AGT   automated guideway transit (1.2)
AGTS   automated guideway transit systems (6.3)
AHS   automated highway systems (8.3)
AIP   Airport Improvement Program (8.1)
ALRT   automated light rail transit (4.2)
APM   automated people mover (1.2)
APT   automated personal transit (1.3)
APTA   American Public Transportation Association (1.2)
ASCE   American Society of Civil Engineers (3)

ATC   automatic train control (3.4)
ATSM   Automated Transit and Shared Mobility (Front Matters)
ATN   automated transit network (1.3.1)
ATO   automatic train operation (3.4)
ATP   automatic train protection (3.4)
ATS   automated transit systems (1.2)
ATS   automatic train supervision (3.4)
AV   automated vehicle (1.3.2)
AVS   Automated vehicles symposium (front matters)

BAA   British Airport Authority (8.2)
BOT   build operate and transfer (8.3)
BPMT   billion passenger mile travelled (6.3)
BRT   bus rapid transit (1.3)
BTS   Bureau of Transportation Statistics (1.2)
BTRM   billion train revenue miles (6.3)
BUPT   billion unlinked passenger trips (6.3)
BVRM   billion vehicle revenue miles (6.3)

CBTC   communication-based train control (4.1.2)
CBD   central business district (4.3)
CCF   central control facility (3.4)
CEA   cost-effectiveness analysis (6.5)
CES   Consumer Electronics Show (8.2)
CFC   consumer facility charge (8.1)
CCVS   computer-controlled vehicle system (2.1)
CL   Circle Line (4.2)
CVG   Cincinnati/Northern Kentucky International Airport (5.1.1)

DB   design build (8.3)
DBFO   design, build, finance, and operate (8.3)
DBO   design-build-operation (8.3)
DBOM   design, build, operate and maintain
DBOT   design-build-operate-transfer (8.3)
DC   direct current (3.2)

DC   destination choice (7.2)
DFW   Dallas–Fort Worth International Airport (4.5)
DLB   driverless bus (1.3.2)
DLLRT   driverless LRT (4.2)
DLM   driverless metro (1.2)
DPM   downtown people mover (2.2)

EIS   environmental impact statement (7)

FAA   Federal Aviation Administration (8.1)
FRA   Federal Railroad Administration (3.0)
FRR   farebox recovery ratio (8.1)
FPS   feet per second (3.1)
FPS^2 feet per second/second (5.1)
FTA   Federal Transit Administration (1.3.2)

GAO   General Accounting Office (1.3.1.2)
GN   guideway network (7.3)
GO   general obligation (8.2)
GPS   Global Positioning System (1.1)
GRT   group rapid transit (1.2)

HR   heavy rail (6.3)

IEC   International Electrotechnical Commission (1.3.1)
IVTT   in vehicle travel time (7.2)

JFK   John F. Kennedy (4.6)

KMPH   kilometers per hour

LCC   life cycle cost (5.3.3)
LHR   London Heathrow Airport (2.4.2)

LIMs   linear induction motors (3.3)
LIRR   Long Island Railroad (4.6)
LPA   locally preferred alternatives (7.1.3)
LRT   light rail transit (6.3)
LRTP   long range transportation planning (7.2)
LVM   Las Vegas Monorail (4.4)

MAC   major activity centers (2.3)
MDBF   mean distance between failures (6.2)
MDT   Miami-Dade Transit (6.3)
MG   Monorail and Automated Guideway (6.1)
MPH   miles per hour (1.3.1)
MPMT   million passenger miles travelled (6.3)
MPO   metropolitan planning organization (7.2)
MR   monorail (6.1)
MRT   mass rapid transit (4.2)
MSF   maintenance and storage facility (3.6)
MTA   Maryland Transit Administration (8.1)
MTBF   mean time between failures (6.2)
MTKM   million train kilometers (6.2)
MTRM   million train revenue miles (6.3)
MTTR   mean time to repair (6.2)
MUPT   million unlinked passenger trips (6.3)
MVRM   million vehicle revenue miles (6.3)

NEL   North East Line (4.2)
NEPA   National Environmental Policy Act (7)
NFPC   National Fire Protection Code (3.4)
NHTSA   National Highway Traffic Safety Administration (1.1)
NJ TRANSIT   New Jersey Transit (8.1)
NTD   National Transit Database (2.3)
NTSB   National Transportation Safety Board (3.0)

OAK   Oakland International Airport (3.2)
OCC   operational control centers (4.1.2)

O-D   origin and destination (5.2.3)
O&M   operation and maintenance (3.4)
OVTT   out-of-vehicle travel times (6.2)

PA   public address (3.1)
PANYNJ Port Authority of New York and New Jersey (4.6)
PATH   Partners for Advanced Transit and Highway (1.3)
PCM   passenger car miles (6.1)
PFC   passenger facility charge (8.1)
PFI   private finance initiative (8.3)
PHX   Phoenix International Airport (3.6)
PLMT   place miles traveled (6.5)
PMT   passenger miles travelled (6.1)
PPHPD passengers per hour per direction (5.2.3)
PPP   public-private partnerships (8.3)
PRT   personal rapid transit (1.2)

RATP   Regie Autonome Des Transports Parisiens (4.1.1)
R & D research and development (8.3)
RER   Reseau Express Regional (4.1.1)
ROI   return on investment (8.3)
RP   revealed preference (10.3)
RTA   Regional Transit Authority (2.4.2)

SAE   Society of Automotive Engineers (1.1)
SAV   shared autonomous vehicle (7.3)
SBT   Singapore Bus Transit (4.2)
SEA   Seattle–Tacoma International Airport (6.2)
SFO   San Francisco (3.2)
SMRT   Singapore Mass Rapid Transit (4.2)
SOV   single occupancy vehicle (3.0)
SP   stated preference (10.3)
SS&PS Systems Safety and Passenger Security (6.3)

TIP   transportation improvement programs (7.2)
TNC   transportation network companies (8.3)
TOD   transit oriented development (9.2)
TPA   Tampa International Airport (6.2)
TRB   Transportation Research Board (1.2)
TRM   train revenue miles (6.1)
TVMs   ticket vending machines (4.4)
SEA   Tacoma International Airport (6.2)

UAACC   user allocation of annualized capital cost (6.5)
UK   United Kingdom (8.2)
ULTRA   urban light transit (2.4.2)
UITP   Union International de Tramways (6.3)
UMTA   Urban Mass Transportation Administration (2.2)
UPT   unlinked passenger trips (6.1)
USDOT   United States Department of Transportation (7.1)
UTA   Utah Transit Authority (8.1)

VAA   vehicle assist and automation (1.3.2)
VAL   vehicle automated léger (automatic light vehicle) (2.2)
VMT   vehicle miles travelled (6.1)
VRM   vehicle revenue miles (6.1)

CHAPTER 1

INTRODUCTION

A few recent developments, such as Google's driverless cars, automated features on various luxury automobile models, and disputes between automobile manufacturers and telecommunication providers on broadband channels, propelled "automated transportation" onto the center stage. It seems that the rapid development in autonomous driving and its control and communication technologies in conjunction with exponential growth in smartphones, detection technologies, and precision mapping and navigation systems will usher in a brand new paradigm: a new fleet of automated or driverless vehicles.

The automated vehicle, be it passenger car or bus, will not only have the potential to change the way we travel, but also the fundamental structures of auto ownership, housing design, and societal relationships. For example, when using a driverless vehicle, there is no need to keep it parked next to downtown offices with expensive parking. A traveler can simply send his or her driverless car to park itself at a remote, cheaper location or home. When driverless cars and self-parking features become a reality, someone may ask why I need a private car. Isn't it easier to summon a driverless car just when I need it? Or why do I need a garage attached to my house if I do not even need to own a car in the first place?

Ready or not, autonomous driving is coming, maybe sooner than many have expected. As transportation planners, engineers, and decision makers are in the process of evaluating technologies, testing prototype vehicles, and

Automated Transit: Planning, Operations, and Applications, First Edition. Rongfang (Rachel) Liu.
Copyright © 2017 by The Institute of Electrical and Electronic Engineers, Inc. Published 2017 by John Wiley & Sons, Inc.

developing safety regulations; the general public may get excited, curious, or sometimes anxious. Capturing the momentum of recent critical development in automated transportation systems, this book provides a comprehensive and systematic evaluation of automated transit technology and its applications, which may lend some significant lessons for the development of automated vehicles.

Building on the extensive research accumulated from more than half of a century and expansive communications with a large network of professionals, this book not only presents a comprehensive review of automated transit technology and development, it also tries to assess existing and potential automated transit applications. As the foundations of in-depth discussions and comprehension, a clear definition of automated transit technologies and their applications is in order and presented in Sections 1.1 to 1.3.

1.1 AUTOMATED TRANSPORTATION

According to the National Highway Traffic Safety Administration (NHTSA, 2013), automated vehicles are those in which at least some aspects of a safety-critical control function, such as steering, throttle, or braking, occur without direct driver input. Automated vehicles may use some combination of onboard sensors, cameras, Global Positioning Systems (GPS), and telecommunications to obtain information in order to make their own judgments regarding safety-critical situations and act appropriately by effectuating control at some level.

NHTSA also clearly excludes "vehicles that provide safety warnings to drivers, such as forward crash warning but do not perform a control function" from fully automated vehicle categories. As shown in Table 1.1, a five-level definition of vehicle automation is included in the policy statement by NHTSA, which provides general guidelines for recognizing automation development.

Around the same time, the Society of Automotive Engineers International (SAE, 2014) has developed its own six-level automated driving classifications. Despite the six versus five levels by SAE and NHTSA respectively, the taxonomy is almost identical except the highest level of automation by SAE. As demonstrated in Figure 1.1, the "Level Zero" automation by either NHTSA or SAE means no automation or only warning but no control function, which maybe equivalent to most of manually driven vehicles on the market as of 2016. On the other hand, the "Level Five," Full Automation by SAE, mandates "full time performance by an automated driving system for all aspects of the dynamic driving task under all roadway and environmental conditions," which is an extremely tall order that is beyond the capabilities of

TABLE 1.1   Highlights of Vehicle Automation Definition

Levels	Extend of Automation	Driver's Role	Automation's Role	Examples
0	No automation	Complete and sole control	Only warning but no control	Collision warning
1	Function-specific automation	Overall and sole control but may cede limited authority	Assist or augment driver's operations	Cruise control, lane keeping
2	Combined function automation	May disengage from actual driving	At least two primary control functions	Adaptive control combined with lane centering
3	Limited self-driving automation	Available for occasional control	No need for driver's continuous monitoring	Automated driving in pre-mapped roadway segments
4	Full self-driving automation	Not needed	All safety critical driving functions and roadway conditions	Empty moving vehicles

Source: NHTSA, 2013.

conventional vehicles as a transportation mode. For example, the "all roadway and environmental conditions" maybe easily construed as there is no paved roadways needed or all of water, land, and air paths can be navigated by such automated machines. Therefore, the NHTSA automation definition of five levels is adopted for the discussion in this manuscript.

Focusing on the four levels of automation, from Automation Level One to Automation Level Four, defined by NHTSA and SAE, it is generally agreed that the vehicle automation definition not only defines the characteristics

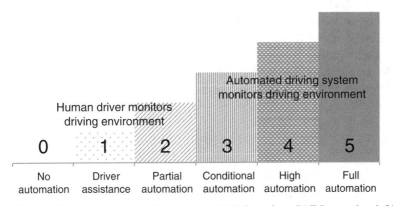

FIGURE 1.1   Vehicle Automation Definition by SAE. Data from SAE International, 2014.

of various levels of automation, but also highlights the developing stages of various automation technologies and their implications to the driving population.

On the other hand, some opposes the numerical level definition of vehicle automation. They believe that the numerical levels suggest an ordering or hierarchy to technology development (Templeton, 2015), which may not be rigidly followed by technology development processes. For example, the Google car is capable of fully automated driving on highways and many private streets approved and mapped by Google at full road speed, which is the first implementation of the "Automation Level Three" concept. At the same time, the "Induct" by French company Navia can be summoned with a phone, drives on ordinary roads among pedestrians, cyclists, and other vehicles, but with very low speed. Without a steering wheel, the "Induct" is fully automated, equivalent to the "Automation Level Four" definition, but the very low speed will not allow it to operate on public highways with mixed traffic.

As of 2016, when this manuscript was developed, there is no "Automation Level Four" vehicle traveling along the highways around the world. However, there are fully automated, driverless, and/or centrally controlled transit vehicles in operation for more than four decades. The millions of vehicle and passenger miles logged by those automated transit applications without a single casualty should be one, but not the only, reason for the automated transportation community to turn our attention to the automated transit applications and learn from their operating experiences.

1.2 AUTOMATED TRANSIT

When hearing the term automated transit, most people would have the images of transit vehicles that do not have drivers in the front, such as the automated people mover (APM) shuttles between airport terminals or monorail trains connecting various casinos in Las Vegas, as shown in Figure 1.2. There are many sizes and shapes of automated transit applications constructed in various public and private locations, but there is no systematic or consensus definition for automated transit except the recent attempt by AP040: Automated Transit System (ATS) Committee, Transportation Research Board (TRB). Defining the research scope, the definition by TRB AP040 only listed all the members of automated transit family (TRB, 2013).

Tailoring definitions of transportation and transit systems by predominate transportation research entities such as Bureau of Transportation Statistics (BTS, 1996) and American Public Transportation Association (APTA, 1994), the author would like to define automated transit as **passenger transportation**

(a)

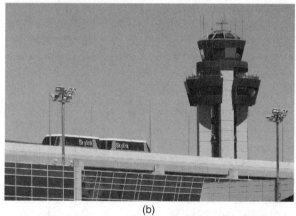

(b)

(c)

FIGURE 1.2   (a–c) Various Images of Automated Transit Systems. *Source*: Liu and Moini, 2015.

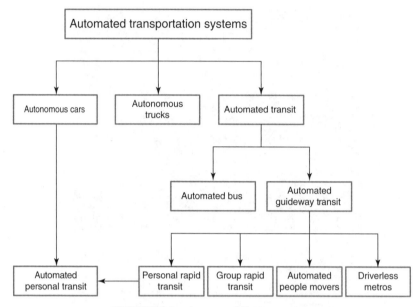

FIGURE 1.3 Automated Transportation.

services that are available to any person who pays a prescribed fare but are not required to be operated by driver, conductor, or station attendant. In practice, the automated transit fare may come from different sources and in various shapes or sizes. For example, the fare for Miami Metromover is zero, or free, and the fare for Morgantown Group Rapid Transit (GRT) by the University students are collected via their tuition. Similarly, the fare for Ultra Personal Rapid Transit (PRT) in Heathrow International Airport in London is included in the parking fees.

Under the general umbrella of Automated Transportation, automated transit is parallel to automated cars and automated trucks, which have no commercial applications at the writing of this book. As shown in Figure 1.3, automated transit is made of a family of individual automated transit modes, such as automated bus, driverless metro (DLM), APM, GRT, and PRT. All of the existing commercial applications belong to the automated transit, especially the automated guideway transit (AGT) group.

As expected, automated transit is different from traditional heavy, light, and commuter rail transit, in that it is operated via a central control system without drivers, conductors, or station attendants (Liu, 2010). Improved communication and control technology has enabled fully automated or driverless, fail-safe operations of modern automated transit to satisfy wider ranges of capacity, spatial coverage, and temporal span of transit services. Traditional guideway transit may join the automated transit family if it is operated or capable of operating via a central control system without drivers onboard the vehicles. Even some agencies may choose to have a vehicle or station attendant present for the comfort of customers such as the Docklands Light Railway (Carter, 1986).

It is possible for traditional guideway transit, such as subway or LRT, to be converted into automated transit applications, as in the case of Paris Metro Line 1. In the other spectrum of lower and medium capacity of Automated Transit applications, AGT, such as APM or PRT, uses narrower right-of-ways, lighter tracks, if any, and smaller vehicles than traditional transit applications.

There are various experiments or testing protocols of automated cars, trucks, and buses, denoted as ovals in Figure 1.3, but there is no commercial application as of 2016. There are a few PRT applications in commercial operations but none of them operate to the full characteristics, such as bypassing stations or direct origin and destination travel; therefore, PRT is also denoted with an oval in Figure 1.3.

Another unique mode, automated personal transit (APT), denoted by a circle, is anticipated once the autonomous driving roadway vehicle technology becomes mature. The main characteristics of APT is that it operates as an autonomous vehicle and not restricted to a guideway as PRT does, while the ownership of APT vehicles resides with a public agency or third party other than the individual user or rider. The formation of the unique APT mode will be dictated by the maturity and customer acceptance of automated vehicles and shared mobility. The significant impacts of the APT on travel behavior, urban development, and other societal aspects are anticipated and will be elaborated in the later chapters.

It is not by accident that vehicle automation applications so far have been concentrated in the fixed guideway subgroup. Vehicle automation within a closed guideway or corridor does have its inherent advantages, which generally rely on a known traveling environment and relinquish control to centralized computer systems.

Many real-world AGT applications have been operating for several decades, which not only proved the feasibility of automated transit services, but also provided valuable experiences in automated operations, customer experience, and market responses. This is also one of the reasons for this book to focus on automated transit, which has accumulated many years of operation and maintenance practices with great safety record. It is understood that automated transit, especially AGT, is different from fully automated roadway vehicles, nevertheless, much information and/or many lessons from automated transit operations may be gleaned and made useful to the overall automated transportation development.

1.3 INDIVIDUAL MODES OF AUTOMATED TRANSIT FAMILY

When zooming into the automated transit family tree, you will notice that there are two main branches under the general umbrella of automated transit: automated buses and AGT. There are quite a few sub-modes under AGT, which includes DLM, APM, GRT and PRT. Each of the sub-modes will be briefly defined in the following section and more characteristics and applications will be described in the following chapters.

1.3.1 Automated Guideway Transit

As members of the automated transit family, AGT is defined as a class of transportation modes in which fully automated vehicles operate along dedicated guideways (Liu, 2010). The capacity of the AGT vehicles ranges from 3 or 4 up to 100 passengers. AGT Vehicles are made of single-unit cars or multiple-unit trains. The operating speeds are from 10 to 55 miles per hour (mph), and headways may vary from a few seconds to a few minutes. AGT may be made of a single trunk route, multiple branches, or interconnected networks.

Depending on the vehicle size, capacity, and other operating characteristics, AGT may be categorized into various subgroups, such as DLM, APM, GRT, and PRT. Different operating environments often give AGT applications generic names, such as airport circulators or downtown people movers. Diversified track configurations, propulsion powers, and other technological features impart to AGT other names, such as monorail, duo-rail, and maglev, among others.

Automation in metro or guideway transit systems refers to the process by which the responsibility for operation and management of the trains is transferred from the driver to the train control system (UITP, 2015). There are various degrees of automation, which is defined according to which basic functions of train operation are the responsibility of staff, and which are the responsibility of the system itself.

Similar to the definition of autonomous vehicles by NHTSA presented earlier, there is a consorted classification of AGT by the International Electrotechnical Commission (IEC). The classification of AGT is explained and exhibited diagrammatically in Figure 1.4. Comparing to the definition for vehicle automation presented earlier, the Grade of Automation by IEC for automated transit is analogous to the four levels excluding the initial "Level Zero" or no automation category from the NHTSA and SAE definitions. For example, a "Grade of Automation One" would correspond to on-sight operation, like a tram running in street traffic. A "Grade of Automation Four" would refer to a system in which vehicles are run fully automatically without

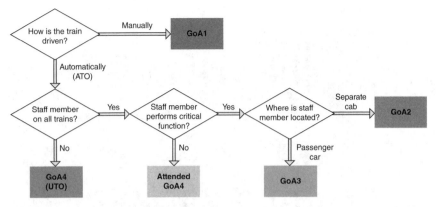

FIGURE 1.4   Illustration of Grade of Automation for Transit. *Source*: Cohen et al., 2015.

any operating staff onboard. The technologies and case studies covered in this book refer to "Grade of Automation Four," which does not require a human driver on board to ensure safety operation of the trains.

In order to accomplish the objective of comprehensive review of automated transit applications and gathering useful lessons learned, the rest of the book will largely focus on the AGT applications that have been in operation and have accumulated sufficient data to be evaluated and compared. Individual modes of those corresponding categories are defined as follows.

1.3.1.1 Driverless Metro   There is no commonly accepted definition for DLM, which often conjures images of heavy rail or subway trains without a driver. Gathering input from various experts and practitioners (Cottrell, 2006; Metro bits, 2015), the author defines DLM as **a metro or subway transit vehicle or unit that operates without onboard intervention from a driver or attendant.**

If the distinction between DLM and traditional transit is whether a driver or attendant is needed to operate the transit service, the feature to separate the DLM from the rest of AGT family lies in the service area it covers and the capacity it provides. In general, DLM operates as a regular fixed route, fixed schedule transit service in high density urban areas. Figure 1.5 depicts the DLM operating in Paris, France.

Besides Paris Metro, which has a high concentration of DLM such as Line 1, Line 14, and Line 15, many metropolitan areas in Asia, Europe, and South America implemented DLMs in recent years. Examples include Dubai Metro, Line 10 in Shanghai Metro, both Line 1 and Line 2 in Copenhagen Metro, and both Line 4 and Line 15 in Sao Paulo Metro. As of 2016, there are nearly 40 DLM lines around the world and more future DLM lines are in the planning and construction stages (Metrobits, 2015).

FIGURE 1.5   Driverless Metro Train for Paris Metro Line 1. Courtesy of Michel Parent.

1.3.1.2 Automated People Movers General Accounting Office (USGAO, 1980) defined automated people movers (APM) as **driverless vehicles operating on a fixed guideway**. This definition did not distinguish APM from DLM or any sub-modes of AGT family, such as GRT or PRT. Using the vehicle capacity yardstick, we can easily distinguish APM, a medium capacity mode of AGT family, from its high capacity DLM, and low capacity PRT cousins.

Building on the GAO basic definition and observing the real-world applications, the APM definition can be supplemented with the following specifications: **APM vehicle, with a capacity ranging from 30 to 100 people, may be operated as single units or as trains at speeds up to 30 mph. APM headway, the time interval between vehicles moving along a main route, varies from 1 to several minutes** (Liu and Lau, 2008).

The APM system is automated in that there are no drivers on board the vehicles or trains. The system is controlled or monitored by operators from a remote central control facility. Typically, the electromechanical design and physical characteristics of an APM are unique and proprietary to each manufacturer (Elliott and Norton, 1999). APM applications may partake names such as downtown people movers (DPM), airport APM (AAPM) or automated trams depending on the operating environments.

FIGURE 1.6   Example of APM Train at Airport.

The very first APM service at a major airport in the United States was installed at Tampa International Airport in 1971 (Lin and Trani, 2000). Today, close to 60 APM applications are seen at various airport facilities worldwide, carrying more than 1.6 million passengers daily (Trans.21, 2014). Figure 1.6 shows the APM at Dallas-Fort Worth International Airport, one of the large-scale Airport APM applications in the world.

1.3.1.3 Personal Rapid Transit   If DLM and APM occupy the large and medium spectrum of vehicle sizes for AGT technology, we can easily place PRT at the other end of the spectrum—very small vehicles with a capacity of three to five persons per car or "pod." Based on definitions in various studies by several authors (Schneider, 1993; Muller, 2007; Koskinen et al., 2007), the author defines PRT as **a subcategory of AGT that offers on-demand, nonstop transportation using small, automated vehicles on a network of dedicated guideways with off-line stations.**

Similar to DLM and APM applications, PRT operates automated vehicles along dedicated guideways. In contrast to DLM and APM applications, PRT vehicles are designed for a single person or a small group traveling together by choice on a network of guideways, and the trip is nonstop with no transfer. PRT stations are often off-line or bypass main lines, so vehicles stop only at their riders' final destination stations. PRT trips typically are on-demand, and PRT vehicles or pod cars are supposed to wait at stations prior to the arrival of passengers. Figure 1.7 exhibits the configuration of a PRT network with stations bypassing the main line guideway, which is the most predominate feature that distinguish PRT from other AGT family members.

FIGURE 1.7 PRT Configuration with Bypassing Stations. *Source*: Zheng and Peeta, 2014. Public domain.

There are currently only a few early, tentative stage applications of PRT, such as Ultra in London Heathrow International Airport, "2getthere" in Masdar City in Abu Dhabi, and "SkyCube" in Suncheon, South Korea. Those applications are labeled as PRTs as the small vehicles do resemble the small, private nature of PRT vehicles, but the network operation and direct origin-destination trips are very limited. For example, Ultra PRT at Heathrow International Airport has only three stations. Trips can and are made between any two without stopping at the third but network maneuvering cannot be tested due to the simple, small-scale application. Both "2getthere" and "Sky-cube" have the by-passing station capability built in but not used.

A fully developed PRT application should serve multiple destinations over a large service area via a variety of paths—a network (Furman et al., 2014). Vehicles will travel from station to station in response to passenger needs and network loads, skipping stations along the way. PRT systems will respond to real-time fluctuations in system capacity by routing vehicles headed to the same destination via the most efficient route available, the routing algorithm may be defined by the dispatching center and change accordingly. Generally, there is no schedule as PRT vehicles typically wait at stations or are dispatched to stations on demand.

There are also different names for PRT applications. For example, automated transit network (ATN) is another name used for PRT in recent years (Furman et al., 2014). In Europe, ATN is often referred as "pod-cars." This book describes the ATN concept and may use it interchangeably with PRT.

1.3.1.4 Group Rapid Transit

After defining the main blocks of the AGT spectrum, DLM, APM, and PRT, it is much easier to envision the middle child—GRT—which is similar to PRT in operating characteristics, but with higher occupancy vehicles and grouping of passengers with potentially different origin–destination pairs. As noted in an early study (National Research Council, 1975), the starting capacity for GRT is six passengers per car, whereas the upper limit is around 16 or 18; there are no clear distinctions between GRT and APM in terms of vehicle capacities. Their differences are inhabited in the areas of station location and vehicle routing patterns.

As the capacity difference blurs between APMs and GRT, it is possible for a GRT to have a range of vehicle sizes to accommodate different passenger loading requirements. That is at different times of the day or on routes with less or more average traffic, a GRT may constitute an "optimal" surface transportation routing solution in terms of balancing trip time and convenience with resource efficiency.

The Morgantown application in West Virginia should be correctly classified as GRT, and it is the only GRT application in the world until 1999 when Rivium GRT started its operation in City of Capelle aan den Ijssel, Netherland (2getthere, 2011). As shown in Figure 1.8, the Morgantown GRT vehicle has seats for eight people and some room for standees. The cars run on rubber tires in a U-shaped concrete guideway that has power and signal rails along the inner walls. The system is fully automated and does not require human drivers. There are three intermediate stations. Each station has several platforms and also "express tracks" that bypass the station completely.

The off-line stations, which can be bypassed for direct origin and destination trips, are one of the distinguished features that tie the Morgantown GRT more closely to PRT systems than their large capacity cousins, such as DLM or APM, which operate more like fixed route and fixed schedule transit. The advantage of the Morgantown application is that it is capable and has been operating among three modes: on demand, circulation, and on schedule. More elaboration on these applications can be found in Chapter 4.

Morgantown succeeded in demonstrating most of the tenets of PRT, but since the vehicles have capacity to hold more than 20 people, which is large in comparison to the PRT concept, transportation theorists frequently refer to the Morgantown system as an example of Automated GRT (Raney and Young, 2005). As a result, we would like to call Morgantown application a GRT.

FIGURE 1.8   GRT in Morgantown, West Virginia. *Source*: Raney and Young, 2005. Public domain.

1.3.2 Automated Bus

Parallel to the definition of bus (United States Government Publishing Office, 2011), an automated bus, or driverless bus (DLB), is defined as **an automated vehicle designed to carry more than 15 passengers and operates on non-exclusive roadways.** As a high capacity autonomous vehicle, automated bus combines the advantages of both driverless technology and high efficiency of public transit. When reaching the level of full automation, an automated bus will operate on non-exclusive roadways, where pedestrian and/or automotive traffic also exists.

There is currently no fully automated bus in commercial operations even as many individual or combined automation functions have been tested in various locations. For example, researchers from California and the Chicago area (Shladover et al., 2004) have tested collision warning, precision docking, transit signal priority, and automatic steering control in Bus Rapid Transit (BRT) applications in the Chicago area. Another group of researchers (Tan et al., 2009) have performed similar demonstration/test of lane guidance function via AC Transit in California.

Another operational use of transit-related Automated Vehicle (AV) technology in the United States is a prototype developed by university under Federal Transit Administration (FTA) grants (National Center for Transit Research,

2015). In Apple Valley, Minnesota, a suburb south of Minneapolis, the Minnesota Valley Transit Authority contracted with the University of Minnesota to develop a GPS-based driver assist system to improve safety during bus shoulder operations.

The latest pilot program funded by the FTA demonstrated the benefits of Vehicle Assist and Automation (VAA) applications for full-size public transit buses in Eugene, Oregon (Liu et al., 2016). The local transit agency, Lane Transit District, contracted with the Partners for Advanced Transit and Highways (PATH) at UC Berkeley, which developed a magnetic guidance system that is used for precision docking by the EmX BRT system at three stations. As shown in Figure 1.9, the automated bus has performed lane keeping, precision docking, and responded to traffic signals along the testing route.

Toyota piloted an automated bus system about a decade ago as a demonstration project during the 2005 Aichi World Expo (Lott, 2014). These robotic buses could electronically couple and uncouple for dynamic platooning of the buses on the fly, and the automated buses could switch between manual and automated operations. The robotic vehicles steer themselves and do not need physical guidance or switches, eliminate the need for costly rail switches.

The latest testing operation of automated bus came from Henan Province in China. In August 2015, Yutong, a leading bus manufacturer in China, has successfully completed its trial operation for driverless or automated bus on the intercity road between Zhengzhou and Kaifeng (Yutong, 2015). Without any human assistance, the automated bus reached the destination safely and reached the top speed 43 mph.

1.3.3 Automated Personal Transit

Propelled by the rapid development of computing, navigation, and communication technologies, vehicle automation is no longer restricted to confined environments or dedicated tracks. Instigated by the ever worsening congestions along our urban street and intercity highways, more travelers are turning their hope for autonomous vehicles, which will liberate human from driving task, an undertaking of 75 billion hours per year in the United States alone (Morgan Stanley Research, 2015). With the widely garnered publicity of Google cars and the like, it is not difficult to image what a great leap or interruption it will be when "Automation Level Four" vehicles become a reality.

According to Morgan Stanley Research (2015), the vehicle automation may very well develop along two diverging paths. As demonstrated in Figure 1.10, the current travel scenario depicted in the first quadrant has been invaded by various shared economy pioneers such as Uber, Lyft, and

(a)

(b)

FIGURE 1.9 (a, b) Automated Bus. Reproduced with permission of Wei-Bin Zhang, UC Berkeley PATH.

Sidecar, which are depicted in the second quadrant. The third quadrant points to the direction of automated vehicles that continue on the current private ownership axis. Far in the future, there will be the convergence of vehicle automation and shared economy—shared autonomy. One of the examples of shared autonomy is the APT defined earlier in this chapter.

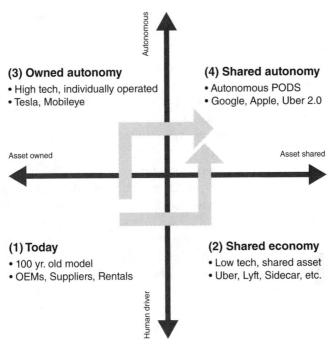

(3) Owned autonomy
• High tech, individually operated
• Tesla, Mobileye

(4) Shared autonomy
• Autonomous PODS
• Google, Apple, Uber 2.0

Asset owned

Asset shared

(1) Today
• 100 yr. old model
• OEMs, Suppliers, Rentals

(2) Shared economy
• Low tech, shared asset
• Uber, Lyft, Sidecar, etc.

FIGURE 1.10   Vehicle Automation and Shared Mobility Paths. *Source*: Morgan Stanley Research, 2015. Reproduced with permission of Morgan Stanley & Co. LLC.

As an integral part of modern life in most developed countries, a private automobile may also be one of the least utilized assess while its expense is only second to housing or shelter. If a vehicle is only utilized one or two hours each day, if the cost of hired taxi can be dramatically reduced via automated vehicles, it is quite possible that individual travelers will forego owning a vehicle all together. It will be much more efficient to summon an automated vehicle when one needs to travel but not have to worry about maintaining, storing, insuring, and owning the vehicle at all. This scenario will usher in a new mode, APT, which combines the advantages of both automated vehicles and PRT. The fleet of APT vehicles will be owned, maintained, and insured by a public agency or third party entity, thus transit mode. It will provide personalized, direct door-to-door service with comfort, convenience, and privacy of an automobile, thus personal.

The fleet of APT vehicles may be owned and operated by a public transit agency, such as New Jersey Transit (NJ TRANSIT), or a private entity, such as Google, an automobile manufacturer, or any third party such as Uber and the likes. An APT vehicle will be liberated from the confined tracks of PRT and expenses of owning a private vehicle. Instead, an APT service will possess some of the characteristics of public transit, accessible to anyone who is willing to pay a fare, operated by a public agency over a regional network. It will also take full advantage of automated vehicles, direct door-to-door services, and reduced cost than taxi since no human driver is needed. The automated or driverless features will keep the cost down and make it affordable for most travelers to hire an automated taxi—another name for APT. The transit classification or public ownership will ensure potential funding sources, regulatory jurisdiction, and safety oversight for the sustainable development of APT.

There are many different shapes and forms of automated transit applications, which maybe called different names depending on their configuration, operating environment, and service characteristics. One common thread connecting the AGT family is that they are operated via central control system without on-board human drivers. Another factual connection is that every individual mode of the AGT family has at least one application in real-world operations. In contrast, the automated bus and APT modes dictated by the non-exclusive roadway operations, are still in various developing and testing stages. More detailed descriptions of individual modes and their respective applications of AGT technologies are elaborated in the following sections of the book.

REFERENCES

2getthere. 2011. "Rivium works." Available at http://www.2getthere.eu/rivium-works/. Accessed in October 2015.

American Public Transportation Association (APTA). 1994. "Glossary of transit terminology." American Public Transportation Association, Washington DC, July 1994.

Bureau of Transportation Statistics, US Department of Transportation. 1996. "Transportation Expressions." Monograph.

Carter, C. 1986. "LRT gives London's Docklands a chance." *Mass Transit*, vol. 13, no. 6, p. 6.

Cohen, J. M., et al. 2015. "Impacts of unattended train operations on productivity and efficiency in metropolitan railways." TRB 94th Annual Meeting Compendium of Papers, Washington DC.

Cottrell, W. 2006. "Moving driverless transit into the mainstream: research issues and challenges." *Transportation Research Record: Journal of the Transportation Research Board*, vol. 1955. ISSN: 0361-1981.

Elliott, D., and J. Norton. 1999. "An introduction to Airport APM systems." *Journal of Advanced Transportation*, vol. 33, no. 1, pp. 35–50.

Furman, B., et al. 2014. "Automated Transit Networks (ATN): a review of the state of the industry and prospects for the future." Mineta Transportation Institute. MTI Report 12-31.

Koskinen, K. R., R. T. Luttinen, and L. Kosonen. 2007. "Developing a microscopic simulator for personal rapid transit systems." In: Proceedings of TRB 86th Annual Meeting Compendium of Papers CD-ROM. Washington DC.

Lin, Y.-D., and A. A. Trani. 2000. "Airport automated people mover systems: analysis with a hybrid computer simulation model." *Transportation Research Record: Journal of the Transportation Research Board,* vol. 1703. ISSN0361-1981.

Liu, R. 2010. "Spectrum of Automated Guideway Transit (AGT) technology and its applications." In: *Handbook of Transportation Engineering,* edited by M. Kutz, McGraw-Hill.

Liu, R. 2016. "Mirage or promised land: automated transit development and lessons learnt for automated vehicles today." In: Proceedings of Transportation Research Board (TRB) 95th Annual Meeting, Washington, DC, January 2016. Peer reviewed.

Liu, R., and C. S. Lau. 2008. "Downtown APM circulator: a potential stimulator for economic development in Newark, New Jersey." In: The First International Symposium on Transportation and Development Innovative Best Practices (TDIBD) by America Society of Civil Engineers (ASCE): Transportation and Development Institute (T&DI), Beijing, China, April 2008.

Liu, R., and N. Moini. 2015. "Proven safety advantages of automated guideway transit systems."In: Proceedings of Transportation Research Board (TRB) 94th Annual Meeting, Washington, DC, January 2015. Peer reviewed.

Liu, R., D. Fagnant, and W. Zhang. 2016. "Beyond single occupancy vehicles: automated transit and shared mobility." In: *Road Vehicle Automation 3,* edited by G. Meyer and S. Beiker, Springer, 2016.

Lott, S. 2014. "Implications for guideway/transit way and station design with respect to automated transit vehicles." In: The 3rd Road Vehicle Automation Conference, San Francisco, 2014.

Metro bits. 2015. "World Metro Database." Available at http://micro.com/metro/table.html. Accessed in August 2015.

Metrobits. 2015. "Driverless metros." Available at http://mic-ro.com/metro/driverless.html. Accessed in August 2015.

Morgan Stanley Research. 2015. "Autonomous cars: the future is now." January 23, 2015. Available at https://www.morganstanley.com/articles/autonomous-cars-the-future-is-now. Accessed in October 2015.

Muller, P. 2007. "A personal rapid transit/airport automated people mover comparison." In: Proceedings of the Fourth International Conference on Automated People Movers IV: Enhancing Values in Major Activity Centers, Irving, TX, March 18–20, 1993.

National Center for Transit Research. 2015. Evaluation of automated vehicle technology for transit. Final Report prepared for Florida Department of Transportation. BDV26977-07. January 2015.

National Highway Traffic Safety Administration (NHTSA). 2013. Preliminary statement of policy concerning automated vehicles. Available at http://www.nhtsa. gov/About+NHTSA/Press+Releases/U.S.+Department+of+Transportation+ Releases+Policy+on+Automated+Vehicle+Development. Accessed in August 2014.

National Research Council. 1975. "Summary of a review of the department of transportation's automated guideway transit program." Monograph. Contract No. OT-OS-40022.

Raney, S., and S. Young. 2005. "Morgantown People Mover – updated description." In: Proceedings of Transportation Research Board Annual Conference, 2005.

SAE International. 2014. "Automated driving: levels of driving automation are defined in New SAE International Standard J3016." Available at http://standards.sae.org/automotive/. Accessed in August 2015.

Schneider, J. 1993. "Designing APM circulator systems for major activity centers: an interactive graphic approach." In: Proceedings of the Fourth International Conference on Automated People Movers IV: Enhancing Values in Major Activity Centers, Irving, TX, March 18–20, 1993.

Shladover, S. E., et al. 2004. "Assessment of the applicability of cooperative vehicle-highway automation systems to bus transit and intermodal freight: case study feasibility analyses in the metropolitan Chicago region." UCB-ITS-PRR-2004-26. California PATH Research Report.

Tan, H-S., et al. 2009. Field demonstration and tests of lane assist/guidance and precision docking technology. UCB-ITS-PRR-2009-12. California PATH Research Report.

Templeton, B. 2015. "A critique of NHTSA and SAE 'levels' of self driving." Available at http://www.templetons.com/brad/robocars/levels.html. Accessed in August 2015.

Trans.21. 2014. "Fall 2014 airport APMs." Available at https://faculty.washington. edu/jbs/itrans/trans21.htm. Accessed in August 2015.

Transportation Research Board. 2013. "Standing Committee on Automated Transit Systems (AP040)." Available at https://www.mytrb.org/CommitteeDetails. aspx?CMTID=116. Accessed in March 2015.

U.S General Accounting Office. 1980. "Better justifications needed for automated people mover demonstration projects." Monograph, CED-80–98.

UITP. 2015. "Automation essentials." Available at http://en.wikipedia.org/ wiki/International_Association_of_Public_Transport. Accessed in March 2015.

United States Government Publishing Office. 2011. "Code of Federal Regulations." Available at http://www.gpo.gov/fdsys/pkg/CFR-2011-title49-vol5/xml/CFR-2011-title49-vol5-part393.xml. Accessed in January 2015.

Yutong. 2015. "Yutong completes world's first trial operation of unmanned bus." Available at http://en.yutong.com/pressmedia/yutongnews/2015/2015IBKCFbteUf.html. Accessed in October 2015.

Zheng, H., and S. Peeta. 2014. "Design of personal rapid transit networks for transit-oriented development cities." USDOT Region V Regional University Transportation Center Final Report.

CHAPTER 2

HISTORICAL DEVELOPMENT

Compared to other transportation modes, such as commuter rail or bus, automated transit is a new kid on the block. The automated transit name has only been used since the Transportation Research Board has formally adopted the same phrase as the name for its AP040 Committee in 2012 (TRB, 2012). However, the concept of automated transit or automated people movers (APM) may be traced back to the sixteenth century. Based on the respective development in terms of concept, technology, and applications around the world as shown in Figure 2.1, automated transit development may be divided into four stages, which are elaborated in this chapter:

1. Conceptual initiations: 1960s and prior
2. Pilot demonstrations: 1970s–1980s
3. Limited applications in confined environment: 1990s–2000s
4. Multipolar development: new millennium and beyond

2.1 CONCEPTUAL INITIATIONS: 1960s AND PRIOR

A romantic, colorful origin of automated transit may be traced back to Salzburg, Austria, the birth place of music giant, Mozart. The early automated

Automated Transit: Planning, Operations, and Applications, First Edition. Rongfang (Rachel) Liu.

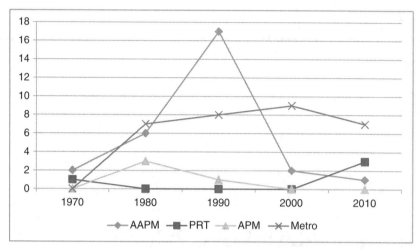

FIGURE 2.1   Automated Transit Applications Worldwide Since the 1970s.

transit configuration in Salzburg was developed in the sixteenth century using a system of water tanks, ropes, and gravity to move vehicles that carried goods up a 625 feet hill with a 67% slope (Juster, 2013). The system is still in use today, but with several modern upgrades. It may be a primitive format of automated transportation as there was no driver in the container/vehicle, but technically it should not be called automated transit or APM as it was primarily used to transport goods.

Despite many versions of how automated transit began, the widely accepted origin of modern automated transit, especially automated guideway transit (AGT) has been documented definitively by Fichter (1964). After a brief review of "metropolis centers" and their associated circulation challenges, Fichter introduced the concept of "individualized automated transit," that is, an automated "small car" operating along "small exclusive traffic ways" within street right-of-ways. With these descriptions and elaborate vehicle control and network layout, a vivid idea of personal rapid transit (PRT) or concept was born in the United States in the 1960s.

Simultaneously, many private citizens in America, such as Edward O. Haltom, William Alden, and Lloyd Berggren, had conceived ideas or developed prototype "Monocab," "StaRRcar," and "Uniflo" vehicles (Anderson, 1996). Several elite universities including Cornell and Massachusetts Institute of Technology (MIT) had embodied automated transit ideas in various design projects or research reports. Quite a few leading research and technology institutions, such as Transportation Technology, Inc. by General Motors, Military Product Group at Honeywell, Inc., and Jet Rail, had designed and/or developed automated transit components or systems. Even some local government staff, such as Robert J. Bartell, Director of Planning for the City of Hartford in Connecticut, had partaken in the development of PRT ideas and connected its impact with urban development.

Looking around the world, you would not have found much initiation on automated transit concept from the other countries in the early stages of automated transit development. The few automated transit activities in foreign countries, such as "Cabtrack" in the United Kingdom, computer-controlled vehicle system (CCVS) in Japan, "Cabinentaxi" in Germany, and "Aramis" in French, were all developed in the late 1960s or early 1970s and directly or indirectly influenced by the initiations from the United States (Anderson, 1996).

It is not by accident that the automated transit concepts grew out of American soil during the 1950s and 1960s as quite a few "catalysts" worked perfectly during that period. First, the concept of automatic control, essential to automated transit, had been firmly established by the early 1950s. Second, the completion of the Apollo Moon Landing Program had freed up government funds and research capabilities and PRT had the potential and promise to fill up the plate. Third, with the fast invasion of automobiles and disappearing streetcar services, some Americans just started to question the validity of automobiles and their far-reaching impact on lifestyle, environment, and society beyond.

The South Park Demonstration Project, "Skybus," in Pittsburgh, PA, was the first attempt in the United States to bring automated transit concept into reality. As an alternative to overcrowding on the city's streets, "Skybus" was a fully automated, rubber-wheeled, electric vehicle that rode on a steel and concrete guideway (Appleby, 2009). As the first automated rubber-tired transit application, the "Skybus" was capable of operating at 60-second headways and a top speed of 50 mph. The "Skybus" vehicles had a capacity of approximate 100 passengers (Sproule, 2001). As demonstrated in Figure 2.2, people waited in long lines during the Allegheny County Fair to ride the modern, thrilling "Skybus." Despite the primitive control and communication technology housed in a telephone booth, rescue vehicle converted from an out of place farm tractor, and many other concerns and unanswered questions, "Skybus" brought the first-hand experiences for more than 30,000 people, who paid a 10-cents fare during the testing period.

The "Skybus" Phase I testing covered 21,000 vehicle miles and was widely considered a success. Several years of testing and modifications proved the automated transit technology reliable, the unique riding experience was attractive and comfortable. However, plagued by opposing attitudes from government agencies - strong support from the Allegheny Port Authority but equally strong opposition from the city and county governments, "Skybus" eventually lost the support of Pennsylvania Governor and public funding. It was abandoned in the early 1970s.

FIGURE 2.2 (a–e) "Skybus" in Pittsburg, PA in the 1960s. *Source*: Appleby, 2009 and Little, 2010.

Looking back, some people may say that "Skybus" was ahead of its time in many aspects. Others may say that the market was simply not ready. However, when looking at the entire development process of the technology or application, it is not difficult to conclude that the ill-fated "Skybus" served a critical function by bringing the automated transit concept from paper to concrete and steel and provided first-hand experiences for many believers and doubters. Among those who visited "Skybus," there was no shortage of industry giants and technology pioneers. Mr. Walt Disney's visit to "Skybus" may be the precursor for many APM installations in Disney Theme Parks. Test rides by executives from Tampa, Newark, and Seattle International Airport played significant roles in their later adoption of airport automated people mover (AAPM) technologies.

2.2 PILOT DEMONSTRATIONS: 1970s–1980s

Automated transit, especially AGT, made great strides starting in the 1970s after many baby steps taken since the 1960s. If individual scholars and private sectors had opened the door to automated transit possibilities, federal funded pilot programs in the United States and other international locations created giant momentum for AGT development.

Burdened by increasing transit operating deficits, traffic congestion, and increasing air pollution problems, the federal government turned its hope to the emerging technology, AGT, a future transportation promise. In 1971, the Urban Mass Transportation Administration (UMTA), the predecessor of the Federal Transit Administration (FTA) today, funded four companies at $1.5 million each to demonstrate its AGT development at a transportation exposition, TRANSPO 72, held at Dulles International Airport near Washington, DC. As a direct result of TRANSPO 72, a few AGT applications were acquired for airports and zoos but not urban transit systems.

Prior to the TRANSPO 72, UMTA had signed a contract with West Virginia University in Morgantown to construct the first AGT application (Schneider, 1993). Having selected the StaRRcar Technology developed by Alden's Self-Transit Systems, a small corporation, UMTA officials decided to add "insurance" by using Jet Propulsion Laboratory, a NASA lab, as the system manager; Boeing, the aircraft giant, as the vehicle manufacturer; and a couple of large engineer firms to design and construct guideways, stations, and other fixed facilities even though none of those companies had any experience or sufficient understanding of the AGT concept.

Propelled by the national mentality that "we can do the difficult today and the impossible tomorrow" that was largely developed immediately after the Apollo Moon Landing, quite a few companies promised that it only took about

2 years to develop an urban demonstration project for AGT. The promises matched well with a political process, that is, if the Morgantown AGT would be ready by October 1972, the President could ride it ahead of election to show case the great technology accomplishments by his administration. With great fanfare to inaugurate the AGT, a new travel mode, no one was interested in hearing the concerns for trial-and-errors of a new technology applications or the slow grinding process of engineering.

The construction of the Morgantown group rapid transit (GRT) began in 1971 and the bulk of the construction was completed indeed 1 year later. However, the extensive testing took much longer so it was opened for passenger service in 1975. As the very first AGT application around the world, Morgantown GRT not only demonstrated the feasibility of automated or driverless transit applications, but also tested the core characteristics of PRT by incorporating bypassing station design in its intermediate stations as shown in Figure 2.3.

According to Lyttle et al. (1986), the purpose of the Advanced Group Rapid Transit (AGRT) Program was to develop an advanced AGT capable of providing high passenger volumes, short waiting times, and high levels of passenger service. The Morgantown GRT consisted of automated vehicles operating on a single lane guideway at short headways with unmanned, off-line stations. The pilot program focused on the critical technologies required to safely command and control the movement of unmanned vehicles along a guideway. However, the vehicles used in the Morgantown GRT, with a capacity of 21 persons including both seated and standees, were much larger

FIGURE 2.3 Morgantown GRT. *Source*: Bell, 2003. Reproduced with permission of Jon Bell.

than the original PRT or "Podcars" design with four or five persons. The much larger vehicles in turn required much wider and stronger guideways and other related facilities. In retrospect, it is believed that the enlarged vehicles and enforced guideways not only significantly increased the capital and operation and maintenance costs but also fundamentally alter the characteristics of the PRT application. Due to the larger vehicle size and small number of stations, five including both terminals, Morgantown GRT had very limited opportunity to test the direct travel from origin to destination with offline stations, the true characteristics of PRT.

Riding high on the waves of AGT promises, UMTA announced its Downtown People Mover (DPM) Program in 1975 and sponsored a nationwide competition among cities (General Accounting Office, 1980). The UMTA DPM Program offered federal funds for the planning, design, and building of AGT as part of the demonstration program. Motivated by the "free" money, the response was almost overwhelming. In 1976, after receiving and reviewing 68 letters of interest and 35 full proposals and making on-site inspections of the top 15 cities, the UMTA selected Los Angeles, St. Paul, Cleveland, and Houston as candidates to develop DPM applications. As second-tier backup candidates, Miami, Detroit, and Baltimore were selected to develop DPMs if they could do so with existing grant commitments. Pressured by the House of Representatives and the Senate Appropriations Conference Committee, the UMTA included Indianapolis, Jacksonville, and St. Louis on the backup candidate list.

After many rounds of debates and discussions, most of the DPM selectees later withdrew from the program, but Miami, Detroit, and Jacksonville stayed the course and inaugurated their AGT services in 1986, 1987, and 1989, respectively. Figure 2.4 exhibits the AGT train operated in Jacksonville, FL, under the DPM Pilot Program by UMTA.

Looking back, few would regard the UMTA's DPM program as a "success." Among all the three cities that implemented DPMs, Miami was often criticized for its higher initial unit costs. However, a recent examination (Cottrell, 2006) indicated that its ridership and costs closely match the original forecast, as shown in Table 2.1. The close match in ridership is usually accomplished after the network was expanded to connect with other transit networks as originally planned, but implemented at a later stage. Overall the DPM program in the United States was only a brief chapter as there was no more DPM application except those three pilot projects. A detailed assessment of those applications is included in the later chapters of this book.

While DPM and PRT development has been riding the roller coaster of novelty thrills, government support, and disappointing implementations in the United States, AGT applications have quietly gained momentum overseas. The initial concept of a fully automated, integrated transit system in Lille,

FIGURE 2.4   Downtown People Mover in Jacksonville, FL. *Source*: Pineda, 1997. Open access.

France was conceived in 1971, almost at the same time that the UMTA initiated its DPM Program. Surprisingly, Lille's vehicle automated léger (VAL) system has run at a profit since 1989. Despite vandalism and concerns over personal safety, ridership figures remain healthy.

The construction for the Lille Metro started in 1978, and the first line was inaugurated in 1983 (Landor Publishing Limited, 1992). When the entire 13.5 kilometers of Lille metro line 1 was opened in 1985, the driverless transit

TABLE 2.1   **Highlights of Downtown People Mover Systems**

Name	Miami Metro Mover	Detroit People Mover	Jacksonville Skyway
Location	Miami, FL	Detroit, MI	Jacksonville, FL
Manufacturer	AEG Westinghouse	YTDC Bombardier	MATRA and Bombardier
Operator	Miami-Dade Transit	Detroit Transportation Corporation	Jacksonville Transportation Authority
Year opened/ expanded	1986/1994	1987	1989/1999
Guideway length (miles)	4.4	2.9	2.5
Number of stations	21	13	3
Fare ($)	Free	$ 0.75	$ 0.5/Free*
2008 Ridership (million)	8.8	2.3	0.5

Source: Sproule and Leder, 2013. Updated by author, 2015.

FIGURE 2.5   VAL Automated Transit Network in Lille, France. *Source*: info@mapa-metro.com, 2010. Public domain.

system linked 18 stations and operated between 5 a.m. and midnight with 1.5- to 4-minute headways. Today, the DLM in Lille covers an impressive 60 stations, expanding from Lille north toward the border of Belgium, as shown in Figure 2.5.

Not coincidentally, a full AGT application was initiated by our northern neighbor in Vancouver, Canada, in the mid-1980s. The "SkyTrain" in Vancouver has three branch lines at the end of 2009: the Expo, Millennium, and Canada lines. The Expo line opened in late 1985 in time for the Expo 86 World's Fair (Castells, 2011); the Millennium line opened in 2002; and the latest, the Canada line, opened in 2009 just in time for the 2010 Winter Olympics. Together, the three branches of "SkyTrain" cover almost 60 miles of track that connects nearly 50 stations. It provides easy and convenient access to Vancouver International Airport and two international border

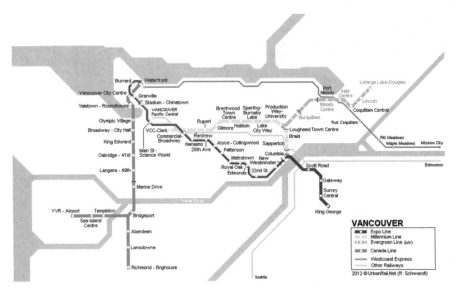

FIGURE 2.6   "Skytrain" in Vancouver, Canada. *Source*: Urban Rail Net, 2009. Public domain.

crossings. Although most of the system is elevated, hence the name "Sky-Train," it runs as a subway through downtown Vancouver and a short stretch in New Westminster, as shown in Figure 2.6.

2.3   APPLICATIONS IN CONFINED ENVIRONMENTS: 1990s–2000s

As the AGT demonstration projects in the urban area faced their continuous criticism due to high cost, low ridership, and most importantly unmet expectations, AGT applications in various airports, major activity centers (MAC), and private institutions, such as amusement parks, hospitals, and museums, have been gaining steam quietly and successfully.

Although the birth of APM occurred in the 1970s, a blossom period emerged since the 1990s when a large number of airport APM applications around the world were established. As shown in Figure 2.7, after a long time period with only a couple of APM applications during the 1970s and 1980s, an increasing number of them have been installed since 1990 and the new millennium.

According to Lea + Elliott (2010), the advent of the U.S. Airline Deregulation Act of 1978 drove airport passenger volumes much higher, which in turn demanded more or larger airport terminal facilities. The emergence of discount airlines in the early 1980s not only fueled the growth of air travel,

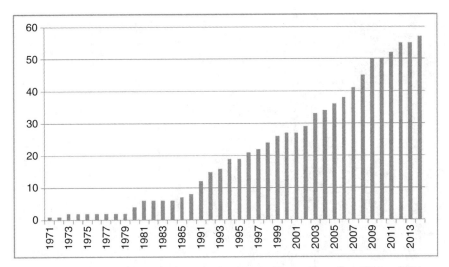

FIGURE 2.7   Accumulated APM Applications in Airport. Data from Fabian, 2014.

but also changed the airline service from point-to-point to hub-and-spoke, which created the need for transfers between airport terminals, especially in large, hub airports. Given the tight foot print of airport terminals and complexity of airport operations, the much improved APM technology with integrated circuits controlling smaller vehicles along compact and light weight guideways lent itself well to connect air passengers between terminals or navigate large airports.

The new expansion of Tampa International Airport in the 1960s designed the central processing facility surrounded on all sides by satellite concourses housing the aircraft gates. Locating the parking garages adjacent to the central processing facility allowed easy access for passengers and also pushed the satellite concourse further from the central processing facility. The distances between various concourses, parking garages, and central processing facility can easily exceed 600 feet, a self-imposed limit for passenger walking distances by the Tampa International Airport executives and also a widely accepted threshold for passenger walking especially when carrying luggage. All of the conventional modes are rejected based on various reasons: moving walkways due to the limited distance coverage and limited throughput, standard light or heavy rail due to their longer headways, large tunnels or elevated track structure, and bus due to multiple steps in boarding and alighting the vehicles and potential interference with aircraft taxi lanes. The airport officials naturally turned their attention to the APM shuttle for Tampa International Airport. Fresh from their memories of the demonstration ride on the "Skybus" in Pittsburgh, PA, the Tampa International Airport executives

FIGURE 2.8   An Early APM Vehicle in Dallas-Fort Worth International Airport. Reproduced with permission of DFW Airport Board.

offered a home for the APM technology developed by Westinghouse in sunny Florida. Figure 2.8 demonstrates one of the early airport APM trains in DFW International Airport. Diagonally across America, the offspring of Westinghouse APM technology found another home in Seattle-Tacoma International Airport outside Seattle, Washington.

After the pioneering development at Tampa and Seattle International Airport, the Airport APM applications flourished in airports around the globe. From the large airside APM in Capital International Airport in Beijing, China (2008), to the planned El Dorado Columbia International Airport APM, there are almost 60 APM applications operating in airports across all five continents (Trans.21, 2014; Little, 2010; Liu and Huang, 2010) as shown in Figure 2.9.

At the same time, various private institutions, such as amusement parks, museums, or hospitals, also host a number of APM applications. According to a recent tally (Liu and Huang, 2010), about one third of people mover applications are located in private institutions as exhibited in Figure 2.10. While there are a large number of institutional AGT applications around the world, there is little information on the historical background owing to the small scale and private nature of the projects; therefore, this book focuses on the development of public or government-funded AGT applications.

Around the new millennium, another unique form of automated transit technology, the monorail, also found its incarnation in various locations. As

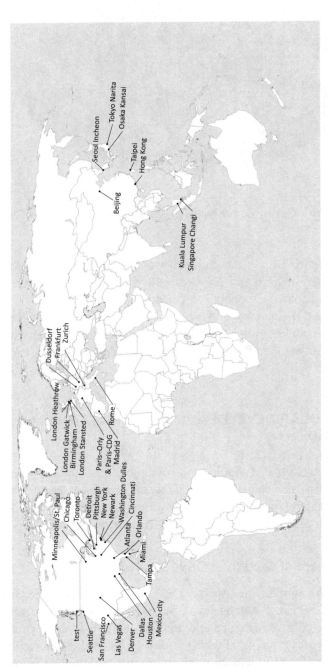

FIGURE 2.9 APM Applications in Airports Worldwide. *Source*: Little, 2010; Liu and Huang, 2010; and Trans.21, 2014.

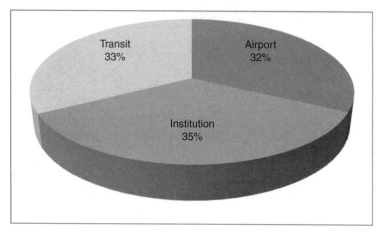

FIGURE 2.10 Distribution of APM Applications. *Source*: Liu and Huang, 2010.

exhibited in Figure 2.11, Kuala Lumpur Monorail and Las Vegas Monorail were inaugurated in 1996 and 2004 respectively. Despite the differences in names and appearances, monorail is considered as a member of the AGT family as defined in last chapter. Its operation and safety records have been collected and included in the National Transit Database (NTD) with other automated transit applications in the United States (Federal Transit Administration, 2012).

2.4 MULTIPOLAR DEVELOPMENT: NEW MILLENNIUM AND BEYOND

As medium capacity APM shuttles and circulators have gradually populated international airports around the world, their counterparts in the large and very small end of capacity spectrum, namely driverless metro (DLM) and PRT have also found their applications in various urban environments and airports.

2.4.1 Exponential Growth of Driverless Metros

The early DLM applications, such as VAL in Lille French; "Skytrain" in Vancouver, Canada; and Kelang Jaya Line in Kuala Lumpur, Malaysia, had created a scattered geographic pattern and thin timeline of automated transit implementations. But the rapid succession of DLM applications around and after the new millennium has certainly filled the voids quickly. For example, in addition to European cities—such as Copenhagen, Denmark (2002); Oeira, Portugal (2004); Turin, Italy (2006); Barcelona, Spain—quite a few

(a) Kuala Lumpur Monorail

(b) Las Vegas Monorail

FIGURE 2.11   Monorail: Another Incarnation of Automated Transit Technology. *Source*: Pixabay. Creative Commons CC0.

large cities in various Asia and South America countries have jumped onto the DLM bandwagon. As demonstrated in Figure 2.12, DLM is no longer confined to Western Europe and Japan. Since the new millennium, quite a few international cities—such as Ankara, Turkey; Bangkok, Thailand; Guangzhou, China; and Sao Paulo, Brazil—have all implemented DLM as part of their respective public transportation systems.

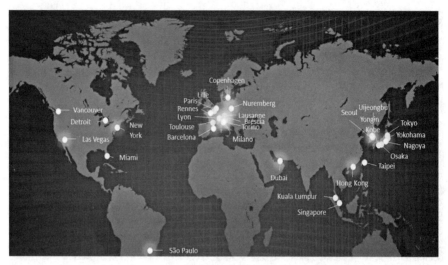

FIGURE 2.12   DLM Applications Around the World. *Source*: Hernandez, 2014. Reproduced with permission of Russell Publishing Limited.

If the very early pioneer of automated transit technology in VAL is considered a lonely experiment with primitive technology, the continuous implementation of DLMs in various French cities such as Lyon (1991), Toulouse (1993), and Paris (1998) has certainly solidified the pioneer position of France in embracing innovation, technology, and converting the most advanced technologies into practical solutions. If there is any doubt about the potential of automated transit and its application in a truly dense urban area or high frequency operation systems, the conversion of Paris Metro No. 1 line, the oldest and second busiest metro line in Paris, from manual operation to driverless in 2011 should have vaporized all those doubts.

As one of the 16 lines that Paris Metro is composed of, the 16.5 kilometer No. 1 line connects La Defense/Grand Arche and Chateau de Vincennes stations, as shown in Figure 2.13. Passing through the heart of the city, Paris Metro No. 1 line is an important east-west transportation route. It transported 213 million travelers since 2008, which equals an average 725,000 riders per day.

It is a pity that there was no automated transit application in the United States since the DPM pilot program in the 1970s and 1980s. However, the recent development in Automated Buses certainly gives us hope: both AC transit in the Bay Area and Eugene, OR have tested level II automated buses equipped with precision docking and lane following features. It is important for transit agencies to test and improve early levels of vehicle automations even though the full automation or automated buses may be far into the future.

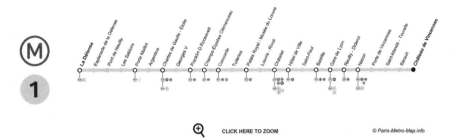

FIGURE 2.13   Paris Metro No. 1 line. *Source*: Pinpin, 2007. Public domain.

On the other hand, it might be puzzling as why there was no AGT application in the United States even though the technology is mature and there are many successful applications in Europe and Asia. Besides the often criticized UMTA pilot program, the author will also explore other factors, such as transit use in general, unions oppositions, safety oversight, and their impact on automated transit applications in the following chapters.

2.4.2 Steady Expansion of APM Systems

The airport APM applications have started to plateau after its rapid growth period during the last two decades. There will still be a few airport APM applications each year while many large and medium-size airports in Asia, South America, and Africa are still in the process of completing and improving their airport environment. However, the overall pace of airport APM installations has become slower since 2010. As shown in Figure 2.14, there was no airport APM application completed in the year 2010, neither in 2013 while there were only two completions in 2014.

Looking through the pipelines of airport expansion or improvement, we did not find a large number of airport APMs in the existing airport plans. However, there are potential expansion territories for airport APMs in developing countries, such as China, India, and Brazil. As the economic conditions have been improving in those countries or regions, travel demand and distance increases, more new airports need to be developed, existing airports may be expanded, or existing conditions of the airport will be improved via APM applications.

2.4.3 Emergence of PRT Applications

As documented in the last chapter, PRT was the prototype conceived by the early pioneers of automated transit development since the 1960s. Fichter

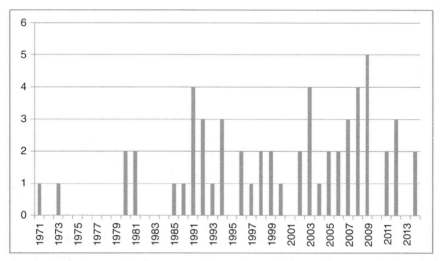

FIGURE 2.14   Airport APM Applications Inaugurated Each Year. *Source*: Trans.21, 2014.

conceptualized the small vehicle, "Veyar," and its extensive network in an urban environment in the 1960s. UMTA attempted the PRT concept in the 1970s in Morgantown, WV, but ended up with a GRT application since it utilizes much larger vehicles, a simpler network, and rarely executed direct origin to destination operations (Office of Technology Assessment, 1975). Despite many criticism and negative publicity, the hybrid Morgantown GRT has been chugging along during the past four decades and more.

It is easy to blame the high cost and primal technology for the isolated application of the Morgantown GRT or the three DPM applications in the United States, but it is hard for the general public or decision makers to pinpoint the critical factors that caused the failure of such demonstration effort. A few academia and true believers of PRT have been trying hard to carry the PRT torch forward. For example, one of the key members of the Economics Evaluation Panel of Automated Guideway Transit in the 1970s (Office of Technology Assessment, 1975), Dr. Edward Anderson, voiced his belief that "personal rapid transit is so promising, and the need for it so imperative, that a significant portion of the federal transit R&D should concentrate on bringing the technology and planning methodology to fruition within the shortest practical time consistent with good manage practice." Dr. Anderson believed that the PRT concept is feasible technologically despite the doubts from other panel members about the possibility of developing dependable, economically feasible PRT within the foreseeable future.

Supported by many believers of PRT and funded by the US DOT research grants, Dr. Anderson organized three International PRT conferences in 1971,

1973, and 1975. Each conference published respective proceedings that addressed a wide range of issues, progresses, and potentials related to PRT. After a prolonged absence of PRT interests and development potentials, another conference series, "Podcar City Conference," was initiated in Uppsala, Sweden, in 2007. The conference location has alternated between Europe and America every year and is in its ninth year in 2015 (Podcar City, 2015).

Dr. Anderson even secured the support of his home institution, University of Minnesota, and established a company, Taxi 2000, to explore the potential for implementing PRT in the late 1980 and early 1990s. Having worked with a number of large engineering firms, such as Stone and Webster and Raytheon, Taxi 2000 was selected by Chicago Regional Transit Authority (RTA) to develop a PRT application. Almost identical to the problems encountered in the Morgantown GRT, Raytheon, another engineering company that converted from defense contractor, had designed a much heavier vehicle, which demanded a guideway twice as wide and deep compared to the original version. The much exaggerated design would certainly have wiped out any promises of small vehicles and resulted in exponential cost increases. As expected, the Chicago RTA has consequently backed out from funding PRT.

Many similar locations, such as Cincinnati, OH (1996); New Jersey (2007); Ithaca, NY (2010); and San Jose, CA (2012), have evaluated the viability of PRT in various urban, regional, and even state-wide applications. Most of the studies gathered information on technology suppliers and related literature and applications, some of them estimated ridership, capital, and/or operation and maintenance costs. Few had advanced to stages of design like Chicago did and none had reached procurement stage.

Similar to the roller-coaster rides of its counterparts in America, many individuals and academic institutions in Europe have carried out their PRT research in isolated locations with or without government support. One particular bright spot is Ultra—urban light transit—a PRT system founded and developed by Dr. Martin Lowson and his designing team. After winning UK National Endowment for Science, Technology and the ARTs (NESTA) funding twice, the Ultra team has developed testing tracks in Cardiff and studied application potential for various European locations.

As the busiest airport in the United Kingdom and Europe and the third busiest airport in the world, London Heathrow Airport (LHR) has five terminals and covers almost five square miles in Western London. In 2005, British Airport Authority (BAA), the operator of Heathrow International Airport, committed to the use of Ultra PRT to provide key connectivity between the new Terminal 5 and a business car park. Given the complex terrain and airport infrastructure—traversing two rivers, seven roads, and a green belt area, the Ultra PRT has to negotiate aircraft surfaces and bridge in-ground services

FIGURE 2.15   Ultra Podcar in Heathrow International Airport in London. *Source*: Lowson, 2010. Public domain.

while conforming to the T5 architecture, which provided a challenging but converted niche for PRT application (Lowson, 2010).

Having collaborated with Arup Inc., a traditional consulting engineering firm, and BAA, Ultra Global Inc. completed the installation of Ultra PRT and inaugurated its service in May 2011 after extensive testing. As the first commercially operational PRT application, the Ultra podcars carry 800 passengers per day as a vital link between T5 business Car Park and the airport terminal. As exhibited in Figure 2.15, there are 21 vehicles running along a 3.8-kilometer one-way guideway, which is dotted by three stations—two in the T5 Business Car Park and one at Terminal 5. In May 2013, the Ultra in Heathrow Airport celebrated a milestone, one million autonomously driven miles (Ultra, 2014).

During the same period, two more PRT applications took place in different Continents, one in Masdar City, Abu Dhabi (Graaf, 2011) and another in Suncheon Bay, South Korea (2014). The five-station PRT with 10 podcars in Masdar City is an urban transit application meant to serve passengers in a carbon neutral, pedestrian-friendly city, where all fossil fuel vehicles are banned. The "Skycube" in Suncheon Bay is more of a shuttle connecting the visitor's center and the bird reserve.

As pointed out by Furman (2014), all three PRT applications in Heathrow, Masdar City, and Suncheon are essentially shuttles and embody PRT functionality to a rather limited extent. There is still a long way to go to reach a network system that will test, prove, or realize the full potential of PRT applications. It seems that the PRT development is currently stuck between a rock and a hard place: there is not enough market interest for a full-fledged system to be implemented while no agency is willing to procure a full PRT system since there is not a proven application.

Will PRT be the next dominating transportation mode of the century? While quite a few dominant "authority figures" were quick to dismiss the PRT idea as "inherently unsound," it might be worthwhile to pause and think since the idea resurges every two decades or so, and there are currently almost two million entries on the Internet that are directly related to PRT.

With an open mind and out-of-the-box vision, some transportation professionals believe that for PRT to become a reality, it may require a revolution in the way we live and travel. That is, PRT may not be feasible if highways and private automobiles continue to be our anchor mode of transportation in the near future. On the other hand, since our society has already spent billions of dollars on building millions of miles of roads and bridges in the past century and has not complained about the expenses, but proudly claimed them as civilization and engineering wonders, it may just be possible to layer PRT guideways on top of the existing roadway networks and replace private automobiles with automated PRT pods.

Others who seek more progressive solutions believe that PRT is capable of adapting to existing living and working patterns, whereas line-haul transit is only efficient in corridor developments. In a large number of metropolitan areas around the world, urban roads are already congested, and land availability and high costs forbid any road expansions. With a much smaller footprint and a fraction of life-cycle costs of conventional transit such as light rail transit (LRT), subway, or commuter rail, PRT may be able to combine the benefits of both private automobiles and public transit by providing a no-wait, well-connected, and origin-to-destination one-seat ride for most urban dwellers.

Practical engineers and rational planners understand that a single mode does not solve all the urban transportation problems; every mode has a place in the mobility spectrum. The applicability is influenced by a variety of factors, such as changing technology, economic conditions, urban development patterns, and social acceptance at particular times. Any entity that is contemplating the idea of PRT or any other form of emerging technologies must undertake systematic research of the respective technology itself and its advantages and disadvantages.

A comparison must be made with other modes, such as GRT, APM, LRT, or automobile, as well as the costs and benefits to users and society at large.

An appropriate viability evaluation, however; should not be confined to technology alone. Market analyses, rider preferences such as mode split given all the travel choices, and cost-benefit analysis also should be part of the viability analysis. Another important aspect of nurturing a technology into fruition is the policy framework that will facilitate its implementation. Potential applications of the technology, engineering specifications, procedural implications, and marketing segmentation should all be examined.

REFERENCES

Anderson, J. M., et al. 2014. "Autonomous Vehicle Technology: a guide for policy-makers." Available at http://www.rand.org/pubs/research_reports/RR443-2.html. Accessed in March 2015.

Anderson, J. E. 1996. Some lessons from the history of Personal Rapid Transit (PRT), 1996. Available at https://faculty.washington.edu/jbs/itrans/history.htm. Accessed in June 2015.

Appleby, E. 2009. "My brief history with Skybus." Available at http://www.brooklineconnection.com/history/Facts/Skybus.html#bm_ed. Accessed in March 2015.

Bauerlein, D. 2013. "JTA will keep skyway free for riders another year." The Flroida Times Union, August 29, 2013. Available at http://jacksonville.com/news/metro/2013-08-29/story/jta-will-keep-skyway-free-riders-another-year. Accessed in December 2015.

Bell, J. 2003. "Morgantown, West Virginia Personal Rapid Transit (PRT)." Available at http://www.jtbell.net/transit/Morgantown/. Accessed in April 2015.

Castells, R. 2011. "Automated metro operation: greater capacity and safer, more efficient transport." *Public Transport International*, vol. 60, no. 6, pp. 15–21.

Cottrell, W. 2006. "Moving driverless transit into the mainstream: research issues and challenges." *Transportation Research Record: Journal of the Transportation Research Board,* vol. 1955. ISSN: 0361-1981.

Federal Transit Administration, 2012. National Transit Database, U.S. Department of Transportation, 2012. http://www.ntdprogram.com. Accessed March 2012.

Fichter, D. 1964. *Individualized Automatic Transit and the City*. Providence, RI. B.H. Sikes, Chicago, IL.

Fichter, D. 1959. "Automated Urban Circulation." MS Thesis, Northwestern University, Chicago, IL.

Furman, B., et al. 2014. Automated Transit Networks (ATN): A review of the state of the industry and prospects for the future. Mineta Transportation Institute. MTI Report 12–31.

General Accounting Office, 1980. "Better justifications needed for Automated People Mover Demonstration Projects." Monograph, CED-80-98.

Graaf, M. 2011. "PRT Vehicle architecture and control in Masdar City." In: Proceedings of Automated People Movers and Transit Systems 2011, America Society of Civil Engineers, pp. 339–348.

Haines, G., et al. 1985. "Otis advanced group rapid transit (AGRT) program. Final report." Monograph for UMTA.

Hernandez, M. 2014. Metro automation: a proven and scalable solution. *Euro Transport*. no. 1, 2014. Available at http://www.eurotransportmagazine.com/13689/supplements/automated-metros-supplement-2014/.

Landor Publishing Limited, 1992. Driverless metros: wave of the future or expensive toy? Urban Transport International, 1992–1995, p. 17.

Info@mapa-metro.com. 2010. "Lille Metro." Available at: http://mapa-metro.com/en/France/Lille/Lille-Metro-map.htm. Accessed in April 2015.

Juster, R. M. 2013. "A trip time comparison of automated guideway transit systems." Master Thesis. University of Maryland, College Park, MD. 2013.

Las Vegas Monorail. 2015. "Convention transportation." Available at http://www.lvmonorail.com/history/. Accessed in February 2015.

Lea + Elliott, Kimley-Horn and Associates, Inc., and Randolph Richardson Associates. 2010. "ACRP Report 37, Guidebook for Planning and Implementing Automated People Movers Systems at Airports." Transportation Research Board, Washington, DC, 2010.

Liu, R., and Y. Deng. 2006. "Research need for personal rapid transit (PRT) and its potential applications." In: Proceedings of the 85th Annual Conference of Transportation Research Board, January 2006.

Liu, R., and Z. Huang. "System efficiency: improving performance measures of automated people mover systems at airports." (10-0835). In: Proceedings of Transportation Research Board (TRB) 89th Annual Meeting, Washington, DC, January 2010. Peer reviewed.

Lowson, M. 2010. "Preparing for PRT operations at Heathrow Airport." In: Proceedings of Transportation Research Board (TRB) 89th Annual Meeting, Washington, DC, January 2010. Peer reviewed.

Lyttle, D. D., D. B. Freitag, and D. H. Christenson. 1986. "Advanced group rapid transit, phase III. Executive summary and final report. Monograph for Urban Mass Transportation Administration.

Malaysian Resources Corporation Berhad (MRCB). 2008. "Kuala Lumpur Sentral." Available at http://www.klsentral.com.my/Conn_Rail.aspx?RAILID=b4abfca9-320f-457c-b853-6667d85b1c60. Accessed in February 2015.

Menichett, D., and T. v. Vuren. 2013. "Considerations for planning and implementing FRT and PRT system sin carbon neutral environments." In: Proceedings of APM-ATs, 2011. ASCE.

National Research Council. 1975. "Summary of a review of the Department of Transportation's Automated Guideway Transit Program." Monograph. Contract No. OT-OS-40022.

National Technical Information Service. 1982a. "Guideway transportation. 1964–1978." Monograph for National Technical Information Service.

National Technical Information Service. 1982b. "Guideway transportation. 1979–1982." Monograph for National Technical Information Service.

Office of Technology Assessment, 1975. "Automated Guideway Transit: An Assessment of PRT and Other New Systems." Prepared at the Request of the Senate Committee on Appropriations Transportation Subcommittee, June 1975. NTIS order # PB-244854.

Pineda, C. C. 1997. "System replacement for the Jacksonville Automated Skyway Express, Phase 1-A." In: Proceedings of the 6th International Conference on Automated People Movers, Las Vegas, NV.

Pinpin. 2007. "Metro Paris M1-plan." Available at http://commons.wikimedia.org/wiki/File:Metro_Paris_M1-plan.svg#mediaviewer/File:Metro_Paris_M1-plan.svg. Accessed in February 2015.

Podcar City. 2015. "Podcar City 9: innovative mobility in the era of automation." Available at www.podcarcity.org. Accessed in December 2015.

Schneider, J. 1993. "Designing APM circulator systems for major activity centers: an interactive graphic approach." In: Proceedings of the Fourth International Conference on Automated People Movers IV: Enhancing Values in Major Activity Centers, Irving, TX, March 18–20, 1993.

Sproule, W., and W. Leder. 2013. "Airport APMs—history and future." In: Proceedings of ASCE 14th International Conference on APM 2013.

Sproule, W. 2001. "Somewhere in time – a history of automated people movers." In: Proceedings of ASCE APM Conference in 2001. San Francisco, CA.

Trans.21. 2014. "Fall 2014 airport APMs." Available at https://faculty.washington.edu/jbs/itrans/trans21.htm. Accessed in August 2015.

Transit Unlimited. 2010. "Las Vegas monorail." Available at http://www.transitunlimited.org/File:Lvmonorail.jpg. Accessed in April 2015.

TRB Committee AP040. 2012. "TRB Committee for Automated Transit Systems." Available at https://sites.google.com/site/trbcommitteeap040/. Accessed in August 2015.

Ultra Global. 2014. "In summer 2011, Heathrow become home to the first commercial Ultra pod system." Available at http://www.ultraglobalprt.com/wheres-it-used/. Accessed in August 2015.

CHAPTER 3

TECHNOLOGY SPECIFICATIONS

Given the long and capricious development process of automated transit technologies and applications, it is critical to sort through voluminous materials to extract accurate and reliable information for potential users. Development and deployment of automated transit has been the goal of a number of individuals and institutions for more than 40 years. Many of them have been driven by the need to find a way to relieve urban congestion while reducing air pollution, minimizing dependence on oil, and reducing or eliminating the need for transit subsidies.

To reduce congestion, many transit advocates simply believe that people should get out of their single occupancy vehicles (SOV) and use public transit. However, the same advocates often scratch their heads and wonder why it is so hard to push or pull people out of their SOVs and lull or tempt them into public transportation systems. After observing the constant tug of wars at many fronts, which include but are not limited to financial, marketing and psychological, it is the author's belief that a true solution is not to pit transit against highway in terms of funding, safety records, or user preferences; the mission is to identify and implement travel modes that are appropriate to the particular segments of travel, which are diverse. Our job is to evaluate and differentiate the characteristics of various automated transit or AGT modes and identify their respective application potentials.

When setting aside the opposing wars between traditional highway and conventional transit travel, it is possible to find such a solution, without

Automated Transit: Planning, Operations, and Applications, First Edition. Rongfang (Rachel) Liu.

prejudice. The new system has to be designed to minimize costs while maximizing ridership and meeting required levels of capacity, safety, reliability, security, and comfort with minimum energy use, less pollution, and integrated land use. A new system has to complement conventional transit systems and make them more effective. During the past half century, automated transit, especially personal rapid transit (PRT), is often anticipated as such a new system or at least possess such potential to be evolved into such new system.

This chapter will present a thorough description of the main components of automated transit. Given the current landscape of automated transit development, the specification focuses on technologies that are in operation stages, such as AGT. There are commonalities among various AGT, such as DLM, APM, and PRT technologies, even though certain unique features exist and separate one submode from another. The unique characteristics of individual submodes are highlighted under each section after introducing the general design or basic prototypes.

There are basically six large categories of AGT components: vehicle, guideway, propulsion and system power, communication and control, station and plat form, and maintenance and storage facilities. One element is covered in an individual section even though there is some overlap when certain elements are integrated or affect each other.

3.1   VEHICLES

AGT vehicles are fully automated, driverless, and either self-propelled or propelled by cables. The vehicle speed, capacity, and maximum train size are usually decided by the types of technologies selected. The typical airport APMs or downtown people movers (DPM) have a capacity of 50 to 75 passengers depending on sitting arrangements and luggage rack characteristics. On the two ends of the capacity spectrum, PRT or podcars usually hold three to six people, whereas the driverless metros (DLM) implemented in large-scale subway systems, such as Paris Metro Line No. 1 and Line No 14, are capable of carrying more than 700 passengers. Figure 3.1 shows the wide spread in vehicle capacity by various AGT applications around the world.

AGT vehicles are usually equipped with thermostatically controlled ventilation and air-conditioning systems, automatically controlled passenger doors, a public address (PA) subsystem, passenger intercom devices, a preprogrammed audio and video message display unit, fire detection and suppression equipment, seats, and passenger handholds. Some APM vehicles, especially in airport applications, are designed to accommodate luggage and luggage carts.

AGT vehicles can be supported by rubber tires, steel wheels, air levitation, or magnetic levitation. Detailed information on each system can be

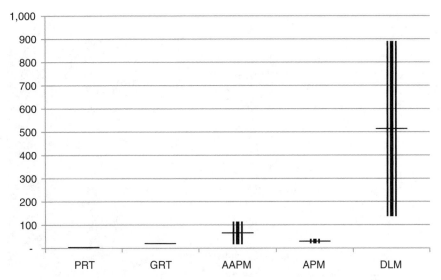

FIGURE 3.1 Vehicle Capacity Range of Various AGT Applications.

found in many engineering books and technical papers (Anderson, 1996; Vuchic, 2007). Vehicle steering and guidance mechanics vary by technology. In general, steering inputs are provided to vehicle bogies through lateral guidance of wheels or similar devices that travel in continuous contact with the guideway—mounted guide beams or rails. The steering inputs cause the bogie, usually located at both ends of each vehicle, to rotate so that vehicle tires do not "scrub" as they move through horizontal curves. Both central and side guidance mechanisms have been used by different manufacturers, and each type has its own unique characteristics.

An example of GRT vehicle used in Morgantown, WV, is shown in Figure 3.2. Detailed dimension and performance specifications for Morgantown GRT are included in Table 3.1. Vehicle features include automatic pneumatic leveling, a cabin heating/cooling system, welded steel frame, emergency exit rear window, impact collapsible front bumper, which withstands 4 feet per second (fps) impact, rigid rear bumper, emergency braking deceleration, and passive power collection from guideway power rails. The front bumper withstands 4 feet per second (fps) impact and rear bumper protects the vehicle control unit. The emergence break is capable of decelerating at 0.062 g at normal operations and 0.45 g at maximum rate. Four-wheel steering is used to achieve 30 foot turning radius. Vehicles steer by means of side guide wheels that sense the location of the left or right guiderail or curb. Switching is accomplished by directing the vehicle to steer either left or right until the guide wheels sense the corresponding guiderail.

Although the Morgantown GRT vehicles are built upon a steerable rubber-tired platform, they do not navigate autonomously. The vehicles are controlled

FIGURE 3.2   Example of GRT Vehicle. *Source*: Bell, 2003. Reproduced with permission of Jon Bell.

via communications from the station control center through the communication loops in the guideway. The vehicles are directed either left or right at the appropriate speed. The original fleet of vehicles from the 1970s is still operating. According to Gannett Fleming (2010), Morgantown GRT vehicles are maintained on a predetermined schedule with 3000-mile intervals. Additional maintenance tasks maybe accomplished at 6000, 9000, and 12,000-mile intervals. The propulsion motor is rebuilt every 90,000 miles. This maintenance schedule essentially rebuilds the vehicle continuously.

Besides its smaller size that caters to individuals or small groups, another unique feature of PRT/GRT is the on-vehicles switch. The switches installed on PRT/GRT vehicles allow real-time route selections by individual

TABLE 3.1   GRT Vehicle Dimension

		Performance	
Length	15' 6"	Control	Automatic remote
Height	8' 9"	Propulsion	70 HP DC motor
Width	6' 8"	Velocity	30 mph / 44 fps max
Weight	8750 lbs empty	Suspension	Air bag automatic leveling
Wheel base	127"	Steering	Side sensitivity (1.2 sec transfer)
Tread width	62"	Brakes	Redundant dual-piston calliper
Capacity	8 Seated, 13 standing	Turning radius	30 feet

Source: Raney and Young, 2005.

vehicles rather than control centers or dispatches. The on-board switches afford PRT/GRT vehicles to shift to optimal routes, especially when AGT networks are in place; but it also increases the complexities of recognition, detection, and navigation functions.

3.2 GUIDEWAY

The guideway is another critical element of AGT applications. It is composed of the track or other running surface, including the supporting structure, which supports, contains, and physically guides AGT vehicles to travel exclusively on the guideway. Guideways can be constructed at grade, underground, or in an elevated structure. Depending on the selected technology, the supplier, and other considerations, the guideway for AGT may be constructed of steel or reinforced concrete. Figure 3.3 shows the elevated guideway for the APM application in Pearson International Airport in Toronto, Canada. The size of the guideway structure varies with span length, train loads, and any applicable seismic requirements. Span length typically ranges from 50 to 120 feet.

The size and weight of the track or guideway can be very different depending on the individual technology and particular applications. For example,

FIGURE 3.3   An Example of APM Guideway Pearson International Airport in Toronto, Canada.

Masdar PRT uses a large single concrete floor showing tire tracks without any other physical guidance mechanism. Vectus applies captive bogey system and Heathrow PRT uses a fairly light guideway strucutre. A very large concrete guideway structure supports the APM application in San Francisco International Airport (SFO) while Oakland International Airport (OAK) Connector builds on a fairly light guideway alignment.

The individual components of an AGT guideway usually include running surfaces, guidance and/or running rails, power-distribution rails, signal rails or antennas, communications rails or antennas, and switches. For technologies that employ linear induction motors (LIMs) for propulsion, guideway equipment may also include reaction rails, also called *rotors*, or the powered element of the motor, also called the *stator*.

Another important segment of guideways is called *crossovers*, and they provide the means for trains to move between guideway lanes. A crossover for most rubber-tired AGT systems is generally composed of two switches, one on each guideway lane, connected by a short length of special track work. Crossover requirements vary significantly among AGT system suppliers, and each supplier's switch and crossover requirements are discrete in that their geometric and other requirements are largely inflexible.

Many AGT guideway configurations have switches that allow trains to switch between parallel guideway lanes or between different routes on a system (Vuchic, 2007). Different AGT technologies have different types of switches, such as rail-like, side beam replacement, and rotary switches. Steel wheel/steel rail AGTs use rail switches, and rubber tire systems, such as the Siemens VAL, use a slot-follower switch that is similar to a traditional rail crossover switch.

As mentioned in the last section, PRT vehicles typically use onboard switching with no moving parts in the guideway. It takes time to activate a guideway switch and more time is needed to verify that the guideway is aligned correctly, and even more time may be needed so that a train can stop if the switch was not verified or verified as not safe. This sequence of switching and routing process dictates that operating headways for guideway switching applications will remain at the "minute" range. With on board configuration, the switch in each PRT/GRT vehicle can be thrown well ahead of the diverging point so it imposes no constraint on vehicle headways.

3.3 PROPULSION AND SYSTEM POWER

Electric power is generally required to propel vehicles and energize system equipment. Propulsion and system power typically are configured such that system operating power will be supplied by power substations spaced along the guideway. The substations house transformers, rectifiers, and the primary and secondary switchgear power-conditioning equipment.

The majority of AGT applications are either self- or cable-propelled. The electrical power for self-propelled AGT vehicles can be either direct current (DC) or alternating current (AC) from a power distribution subsystem, while small vehicles, such as PRT vehicle or podcars, can be powered by an onboard battery, which may be charged along the guideway or traveling route. Cable-propelled vehicles are attached to cables and pulled along the guideway. Some cable systems have vehicles permanently attached, whereas the latest applications allow vehicles to be attached or detached depending on the operation needs.

Self-propelled AGT systems may use electric traction motors or LIMs. Self-propelled APMs are electrically powered by onboard AC or DC motors using either 750/1300 V DC or 480/600 V AC wayside rail-based power-distribution subsystems. Self-propelled AGTs are not limited in guideway length. These technologies can be used for many guideway configurations. Onboard battery propulsions liberate AGT vehicles from the confinement of guideway length or spacing density.

Cable-propelled AGTs use a steel cable or "rope" to pull vehicles along the guideway. The cable is driven from a fixed electric motor driver located along the guideway. These technologies are usually used for shorter shuttle systems, typically for distances up to 4000 feet. Onboard equipment power is usually provided by 480-V AC wayside power. There has been recent interest among airport owners/operators in reducing power requirements for APM systems. Heightened cost awareness, the variable price of energy, and a focus on sustainability have created a strong interest in lowering the power requirements around the airport, including the APM system.

One of the examples of cable-propelled AGT applications is the Toronto Pearson International APM system. The cable wheel drive in the tensioning tower weighs about 75 tons and is powered by two 1500-kW motors. As shown in Figure 3.4, the eight-strand galvanized "rope" runs in one continuous loop within the guideway. A fixed grip assembly forms the mechanical connection between the train and the cable, which is accelerated, decelerated, and stopped by a stationary machine-drive system that propels the cable. Should there be a power failure, an emergency diesel source kicks in to provide power for communication and for heating, ventilating, and air conditioning.

3.4 COMMUNICATIONS AND CONTROL

AGT applications are marked with automated central control and communication systems, which are different from conventional guideway transit, such as commuter rail or subway. All AGT applications use command, control,

FIGURE 3.4   A Sample for Cable Propeller.

and communications equipment to operate the driverless vehicles. Each AGT supplier, based on its unique requirements, provides different components to house the automatic train control (ATC) equipment. ATC functions are accomplished by automatic train protection (ATP), automatic train operation (ATO), and automatic train supervision (ATS) equipment.

ATP functions ensure absolute enforcement of safety criteria and constraints. ATO equipment performs basic operating functions within the safety constraints imposed by the ATP. ATS equipment provides automatic system supervision by central control computers and permits manual interventions and/or overrides by central control operators using control interfaces.

The AGT system includes a communications network monitored and supervised by the central control facility (CCF). The network typically includes a station public address (PA) system, operation and maintenance (O&M) radio systems, emergency telephone, and closed-circuit television. The basis for many of those communication requirements is emergency egress codes such as National Fire Protection Code 130 (NFPC), (2014). As shown in Figure 3.5, the CCF is the focal point of the control system and can vary in size, which may range from a simple room with one or two operator positions with a minimum of computer and CCTV monitor screens to a large room with multiple operator and supervisor positions and a large array of screens and other information devices.

FIGURE 3.5   A Typical Control and Communication Station in CCF.

As the standard requirement for automated transit, safety equipment responds in real time while computer hardware and software performs vital control functions to reduce the risk of human error and ensure the ultimate safety of the system, passengers, and staff (Chaussoy et al., 1999). For example, the moving block collision avoidance system developed for the advanced group rapid transit (AGRT) applications utilizes the onboard vehicular odometer to track individual vehicle speeds and positions throughout the guideway (Colson et al., 1984). The design utilizes microprocessors and incorporates unique self-exercised software to detect potentially unsafe latent failures within the hardware. Odometer data are downlinked to the wayside via an inductive communications system. Wayside equipment computes minimum safe headway for each vehicle based upon a worst-case scenario, and compares this worst-case value to the measured value. Headway violations initiate emergency braking for affected vehicles.

3.5 STATIONS AND PLATFORMS

Similar to conventional transit stations, AGT stations are located along the guideway to provide passenger access to the vehicles. AGT stations generally are equipped with automatic platform doors and dynamic passenger information signs. As *sine qua nons* for any AGT applications, these features have also been implemented in conventional transit, especially those newer and larger systems, as optional improvements. The AGT stations typically have station equipment rooms to house command, control, and communications equipment.

The barrier wall, door sets, and passenger circulation/queuing area within the AGT station and adjacent to the AGT berthing position are commonly referred to as the *platform*. A single station can have multiple platforms. The types of platforms used depend on the type of AGT configuration, physical space constraints, and any passenger-separation requirements. In addition to the automatic train doors, the stations may also have platform doors that align with a stopped train, and the two-door systems operate in tandem. The automatic station platform doors provide a barrier between the passengers and the trains operating on the guideway. These doors are integrated into a platform edge wall, as shown in Figure 3.6.

Similar to conventional transit station platforms, the configuration of platform for AGT applications may include lateral, central, or combination of both. As shown in Figure 3.7, the lateral platform, 3.7a, provides a platform for each direction, which may separate the directional flows easily if desired by transit operation and/or fare collection. On the other hand, central platform may require less investment in infrastructure, such as escalator, elevator, and service facilities.

The station platform doors provide protection and insulation from the noise, heat, and exposed power sources in the guideway. The interface between the station platform and the AGT guideway is defined by the platform edge wall and automated station doors. Dimensions defining the minimum width of AGT platforms and stations are developed based on analyses that take into account train lengths of the design vehicle, reasonable allowance for passenger circulation and queuing at the platform doors and escalators, passenger queuing and circulation requirements based on ridership flow assumptions, and reasonable spatial proportions and other good design practices. Stations for airport APMs typically are "on-line" with all trains stopping at all stations. Detailed analyses of passenger flow and travel patterns are needed to determine the best platform configuration to suit a particular AGT application.

One of the fundamental differences between PRT and other AGT family members is the location and function of the stations. A true PRT system should have "off-line" station locations to enable direct travel between origin and destination stations. Borrowing a phrase from the shipping industry, PRT designers sometimes use "berth" to define the spaces allocated for individual podcars along the platform. Figure 3.8 depicts the off-line station with multiple berths for Morgantown GRT. The Morgantown off-line station was

(a)

(b)

FIGURE 3.6   (a) Platform Doors in AAPM. *Source*: DFW International Airport. Reproduced with permission of Steven E. Poerschmann, DFW International Airport. (b) Platform Doors in DLM Stations. Courtesy of Michel Parent.

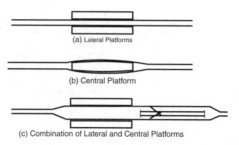

FIGURE 3.7   Platform Configurations. *Source*: Vuchic, 2007. Reproduced with permission of John Wiley & Sons, Inc.

originally designed to allow vehicles moving off the mainline and unload passengers without obstructing other through vehicles on the mainline. However, since the GRT is often operated in shuttle or circulation modes, the off-line station function has not been used to their full potential.

Spatial, temporal, and institutional issues associated with intermodal connections are important for any transit service but are even more predominant in AGT stations. Dynamic passenger information signs typically are installed above the platform doors and/or suspended from the ceiling at the center of the station to assist passengers using the system. These dynamic signs provide information regarding train destinations, door status, and other operations. In PRT applications, additional passenger interfaces are generally needed to summon vehicles for service on demand. Figure 3.9 demonstrates the station configuration in Heathrow PRT.

3.6 MAINTENANCE AND STORAGE FACILITIES

Another key element of the AGT system is the maintenance and storage facility (MSF), which provides a location for vehicle maintenance, storage, and administrative offices. The maintenance function is not limited to maintain, clean, wash, and repair vehicles. It may also include shipping, receiving, and storage of parts, housing of tools, spare equipment and spare vehicles. Some application, such as APM in Phoenix International Airport (PHX), actaully locates its control center in the same vacinity of the MSF.

The MSF is typically located away from the operating alignment in the larger AGT systems, such as airport APM, DPMs, or DLMs. Vehicle testing and other testing track functions are generally performed along the guideway approaching the facility when the MSF is separated from the operating guideway. Simple, smaller shuttle systems often have the MSF located "under" one of the system stations or toward the end of the operating alignment. Figure 3.10 is an example of the MSF located at the end of the operating alignment at Pearson International Airport in Toronto, Canada.

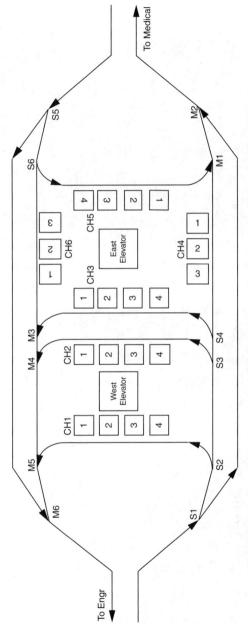

FIGURE 3.8 Bypassing Stations for PRT Applications. *Source:* Raney and Young, 2005. Public domain.

FIGURE 3.9   PRT Station at Heathrow International Airport. Reproduced with permission of John Eddy/Arup.

FIGURE 3.10   Maintenance and Storage Facility in Pearson International Airport, Toronto, Canada.

It is noted that the placement and sizing of the MSF is usually due to site considerations and the size of the vehicle fleet. Within the MSF, there will probably be a few maintenance bays designed for heavy maintenance, which usually requires a pit or vehicle lift for easier access to underside components. There will also be a few bays set aside for light maintainance of vehicles.

Unlike traditional railroad and conventional transit, which are regulated or overseen by the Federal Railroad Administration (FRA) and the National Transportation Safety Board (NTSB), respectively, AGT applications do not have a clear jurisdiction in terms of safety oversight and enforcement. However, since AGT applications are inherently complex systems that involve multiple interacting subsystems, new technology, and public safety, it is essential to establish minimum standards for their design, construction, operation, and maintenance.

Realizing the benefits of standardization to organizations that specify and procure AGT systems, such as regulatory authorities, system suppliers, system operators, system users, and the general public, the American Society of Civil Engineers (ASCE) has taken the lead in developing "Automated People Mover Standards" (Committee of Automated People Mover Standards, 2006). The APM Standards by ASCE include minimum requirements for design, construction, operation, and maintenance of APM systems, especially on the subject of the physical operating environment, system dependability, ATC, and audio and visual communications. The ASCE standards have no legal authority in their own right and have not been officially adopted by any authority that has jurisdiction over AGT applications; nevertheless, the standards serve as a general guideline for transportation professionals to plan, build, and manage AGT systems in the years to come.

REFERENCES

Anderson, J. E. 1996. "Some lessons from the history of Personal Rapid Transit (PRT)." Available at https://faculty.washington.edu/jbs/itrans/history.htm. Accessed in June 2015.

Automated People Mover Standards Committee. 2006. "Automated People Mover Standards, Part 1, ASCE 21-05." Available at http://www.apmstandards.org/SectionPublications/APMStandards1&2.htm. Accessed in July 2015.

Bell, J. 2003. "A car waits for departure at the Walnut Station." Available at http://www.jtbell.net/transit/Morgantown/. Accessed in July 2015.

Chaussoy, J., et al. 1999. "The Importance of Software and Safety Audits for Automatic Metros." Business Briefing: Global mass Transit Systems, World Markets Research Centre.

Colson, C. W., et al. 1984. "Advanced group rapid transit odometer data downlink collision avoidance system design summary. Final report." Monograph for UMTA. UMTA-WA-06-0011-84-2.

Gannett Fleming, Inc. 2010. "PRT Facilities Master Plan." Prepared for West Virginia University. June 2010.

National Fire Protection Association. 2014. "Standard for fixed guideway transit and passenger rail systems." Available at http://www.nfpa.org/codes-and-standards/document-information-pages?mode=code&code=130. Accessed in July 2015.

Raney, S., and S. Young. 2005. "Morgantown people mover – updated description." In: Proceedings of Transportation Research Board Annual Meeting, Washington, DC, January 2005.

Vuchic, V. 2007. *Urban Transit Systems and Technology*. John Wiley & Sons, 2007.

CHAPTER 4

APPLICATIONS

The current status of automated transit and its applications are critical not only for our understanding of the automated transit technology development but also anticipating the market for automated transportation as a whole. As indicated by Shladover (2012), a leader in autonomous vehicle research, "some of the strongest progress in vehicle automation has already been made in the field of transit. Indeed, one could consider the wide variety of airport people movers and automated urban metros to be examples of existing deployed automated vehicles, except these are mechanically captive to their guideways."

Automated transit comes in a variety of applications (Swede, 1992). Some systems roll on rubber tires while others are pneumatically levitated. Both top and bottom supported applications can be found, some propelled with rotary electric motors while others are cable-drawn. A number of real world applications based on diversified DLM, APM, and PRT technologies in various planning, construction, and operation stages may serve as theoretical and practical laboratories for us to examine various aspects of automated transit technologies and their respective successes and failures in meeting the particular travel needs of various markets.

According to a recent tally (Trans.21, 2014), there are more than 100 applications along the continuous spectrums of AGT technologies around the world. After more than four decades of emerging, growing, and maturing, the

Automated Transit: Planning, Operations, and Applications, First Edition. Rongfang (Rachel) Liu.
Copyright © 2017 by The Institute of Electrical and Electronic Engineers, Inc. Published 2017 by John Wiley & Sons, Inc.

AGT technology is no longer limited to airport use as shuttles or circulators. It has expanded to downtown and metropolitan areas as major activity center circulation and public transit systems. A few selected case studies, from high to medium to low vehicle capacity systems, are presented in the following sections.

4.1 DRIVERLESS METRO IN PARIS

The Paris Metro is chosen as the case study of driverless metro (DLM) since it includes not only the oldest and busiest metros in the world but also it showcases both installing a brand new DLM, Line No. 14, and converting a manually driven operation to a fully automated DLM, Line No. 1.

4.1.1 Clean Slate of Automation: Line No. 14

The Eiffel Tower maybe a symbol of Parisians embracing technology and innovation in the nineteenth century, as it tested many advanced technologies during that era (Jonnes, 2010). The Eiffel Tower was surpassed as the tallest building in the world after less than half a century but the French determination to lead the technology and innovation parade was found again in their brave implementation of DLM in Paris toward the end of the twentieth century.

In 1987, the Regie Autonome Des Transports Parisiens (RATP) proposed "Project Météor" to create a new Métro line, from Porte Maillot on the edge of the 16th arrondissement to the *Maison Blanche* district in the 13th, an area poorly served by public transport despite its large population. As shown in Figure 4.1, the line was modified later to actually originate from a terminus at Saint-Lazare, with the possibility of extending to Clichy and assimilating the Asnières branch of No. 13, thus simplifying its complicated operation.

This new Paris Metro Line, No. 14, took the opportunity to incorporate innovations, which served as testing labs for the rest of the network: the stations are larger and, at 120 meters, longer and thus can accommodate eight-car trains. The station spacing are longer, allowing an average speed of close to 40 kilometers per hour (kmph), close to double that of the other lines. Most importantly, the line is completely automated and runs without any driver, which is the first for Paris Metro and also the first major metropolitan line in a national capital city to do so.

As documented in Table 4.1, the first segment of Paris Metro No. 14 was inaugurated in 1989. The middle section, between Chatelet and Gare de Lyon stations, overlaps and provides relief for Reseau Express Regional (RER) Line A, the heavily traveled commuter rail line in Paris (Systra, 2013). The

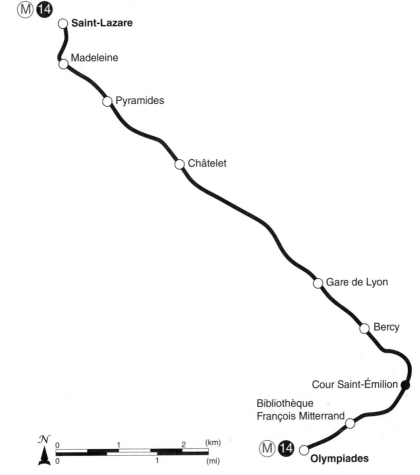

FIGURE 4.1   Route Map for Line No. 14, Paris Metro. *Source*: Motty, 2015. Public domain.

fleet for No. 14 is made up of 52 driverless, rubber tired, MP 89 trains. The first version of rolling stock comprises six-car trains, measuring 90 meters with a passenger capacity of 720. With an average speed of 40 kmph, it only takes 13 minutes to traverse the 8.6 kilometers between Saint-Lazare and Olympiades. As of January 2015, a new train order of 518 million Euros has been confirmed for a fleet of 35 eight-car trains, which will increase the capacity of No. 14 significantly (Brighinshaw, 2007).

Travelers have been largely satisfied with No. 14's speed and services even though it is not free of incidents. As is often the case in a guideway transit, platform doors prevent access to the tracks, making them susceptible to electric outages that may interrupt service. On September 20, 2004, two trains stopped entirely in the tunnel after a signaling failure (Metro-pole, 2011). On

TABLE 4.1   **Key Statistics for Paris Metro Line 14**

Name	Paris Metro Line 14
Location	Paris, France
Operator	RATP
Category	DLM
Configuration	Line
Length (KM)	8.6
Stations	9
Platform doors	Yes
Trains	MP 89 by Alstom
Train capacity	90 meters with six cars
Daily ridership	450,000
Headways	85 seconds
Fleet size	52
Inauguration	1989
Capital cost	1.3 billion Euros

December 22, 2006, passengers were trapped for one and a half hours after an electrical failure on the line that arose from a mechanical failure (Metro-pole, 2011). Other technological failures have occurred twice: on March 21, 2007, traffic was interrupted between Gare de Lyon and Bibliothèque François-Mitterrand (Metro-pole, 2011) and again on August 21, 2007, and both were declared as technical failures.

Despite those minor glitches, traffic on No. 14, the first DLM line, grew quickly. As shown in Figure 4.2, the ridership for the first full year of operation is almost 20 million, it reached 30 million in 2003, just 5 years in service despite the fact that the service was interrupted several times to allow the

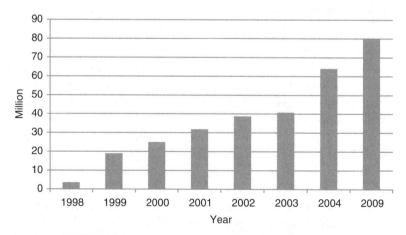

FIGURE 4.2   Ridership Trend Along Paris Metro Line 14.

installation of an extension north from Madeleine to Saint-Lazare. By 2009, the annual ridership reached 80 million, when the full automated line was barely 8 years old. Given the steady growth of ridership on No. 14, it is clear that passengers were undeterred by automation or driverless operations but benefited from the increased capacity through shorter headways and on-time operations.

4.1.2 Conversion from Manual to DLM: Paris Metro Line No. 1

It is not a surprise that the largest and first DLM took place in Paris, which often leads the world in many fronts of technology, fashion, and life style. If the Paris Metro Line No. 1, inaugurated in 1900, represented a pioneer application in the rail public transit arena, the continued commitment to lead automation and innovation of transit operations by the RATP, the transport authority in Paris, was evident in their implementation of automated train control, communication-based train control, and various operation technologies, which eventually led to the full automation, or DLM.

As revealed by its name, Paris Metro No. 1 Line was the first line to open in 1900. Being one of the 16 lines made up the Paris Métro, Line No. 1 connects the La Défense – Grande Arche and Château de Vincennes stations. As shown in Figure 4.3, with a length of 16.5 km, Paris Metro Line No. 1 constitutes an important east-west transportation route for the City of Paris. With an annual ridership of 213 million in 2008 or nearly three quarters of a million daily, the Paris Metro Line No. 1 is the second busiest subway route in Paris (Juillet, 2013).

If the new implementation of driverless operation in Paris Metro Line No. 14 symbolizes the innovation spirit of Parisians, the conversion of Line No. 1 from manual to automatic represented the cultural shift from technology to customer service in the twenty-first century. The idea of automation for Paris Metro Line No. 1 was conceived around 2001 and the initial focus was to improve individual fields of activities, such as transport system, stations, and the commercial side of operations (Mancone, 2011).

A series of events occurring around the new millennium have helped to solidify the conversion plan for Paris Metro Line No. 1 toward driverless operations. First, a serious accident, a driver exceeded the speed limit in the curve for Notre Dame de Lorrete along Line No. 12, resulting in an overturned train, 24 injuries, and significant damage to the train, track and operations. The dramatic event has evoked serious concerns and reminded the metro operators, passengers and other stakeholders of the importance of rail safety, which was later adopted as the number one priority for RATP.

Second, a series of strikes of metro train drivers took place around the beginning of 2001. Having been excluded from the discussion about

Château de Vincennes

Bérault

Saint-Mandé - Tourelle

Porte de Vincennes

Nation

Reuilly - Diderot

Gare de Lyon

Bastille

Saint-Paul

Hôtel de Ville

Châtelet

Louvre - Rivoli

Palais Royal - Musée du Louvre

Tuileries

Concorde

Champs-Élysées Clemenceau

Franklin D. Roosevelt

Georges V

Charles de Gaulle - Étoile

Argentine

Porte Maillot

Les Sablons

Pont de Neuilly

Esplanade de la Défense

La Défense

© Paris-Metro-Map.info

CLICK HERE TO ZOOM

FIGURE 4.3 Route Map for Paris Metro No. 1.

organizational changes, the trade unions were irritated and utilized strike as the only way to express their disagreement and/or get more compensation and benefits for employees. The strikes in the winter of 2001 not only disrupted the commuting services for Parisians but also affected tourist and other economic activities in the capital of France. As a result, it has shaken the key values and core beliefs of RATP, which consequently made "continuity of service" and "quality of service" as second and third highest priority, only next to safety.

Third, by 2001, the driverless operation of Line No. 14 was proven successful after more than 2 years in full service. The success was not only reflected in the safe operation without any major glitches but also highlighted by the increasing ridership; more and more passengers were using the automated line and initial ridership goals were surpassed before its third year operations. If the clean slate of Line No. 14 driverless operation still has the flair of technology breakthrough, the conversion of Line No. 1 from manual to automatic operation was completely motivated by the customer service. The focus is on the "continuity and qualify of service" provided to customers.

The conversion from manually driven operation to fully automated operation for Paris Metro Line No. 1 took place from 2007 to 2011 in a staged process. The conversion not only ushered in new rolling stocks, the MP 05 trains, but also introduced the laying of platform edge doors in all stations, as shown in Figure 4.4.

Similar to the integration of Paris Metro networks, the modernization of Paris Metro Line No. 1 was not isolated but intertwined with the update of the entire Paris Metro system. The automation of Line No. 1 was largely decided because of its operating characteristics: aging rolling stock and control system, busiest segment, and significant variability in the number of passengers due to tourism. Besides the conversion of Paris Metro Line No. 1 to DLM, some Paris Metro Lines were modernized with communication-based train control systems (CBTC) and others with new operational control centers (OCC).

Contrary to the common belief that automation eliminates jobs, particularly the train driver's jobs, Paris Metro has created more attractive and higher paid jobs for metro train drivers via DLM applications. Some of the train drivers were promoted to managers to ensure the safe and continuous operation of Line No. 1. Others were trained and assigned to operate other metro lines, resulting in a larger pool of train drivers with more interchangeable skills and flexible shifts, which becomes the back bone of "continuity and quality of service" for Paris Metro (Systra, 2013).

Automation not only allowed Paris to remain as a model for technological innovation in the railway industry but also permitted an increase in the number of lines in normal service when transit workers are striking, even though there was no notice of strike since the conversion. Although most of the stations remain the same as they were prior to automation with the exception of the

FIGURE 4.4 Berault Station Along Paris Metro Line No. 1.

platform screen doors, many stations, such as St. Paul, received brand new signage. Franklin D. Roosevelt Station received a complete overhaul from its old post-World War II facade to a more contemporary and modern look.

4.2 AUTOMATED LRT IN SINGAPORE

Amalgamating the advantages of Metro, LRT and APM, a series of automated LRT (ALRT) applications have been implemented in various locations around the world. The ALRT or driverless LRT (DLLRT) combines proven technology with two important innovations: linear motors and steerable axle bogies (Parkinson, 1986). Together, those two features afford central control, minimize noise, and reduce wear and tear on the rails laid to the standard gauge. Bukit Panjang LRT line in Singapore, opened in November 1999, is one of the international applications of ALRT and described here.

As a feeder to Mass Rapid Transit (MRT) and bus services in Singapore, the 7.8 kilometer Bukit Panjang ALRT line plays a significant role in accomplishing the national transit access goal for Singapore—establish public transportation access within a quarter mile (400 meters) walking distance of every citizen in the country (Robbins, 1996). As shown in Figure 4.5, The Bukit

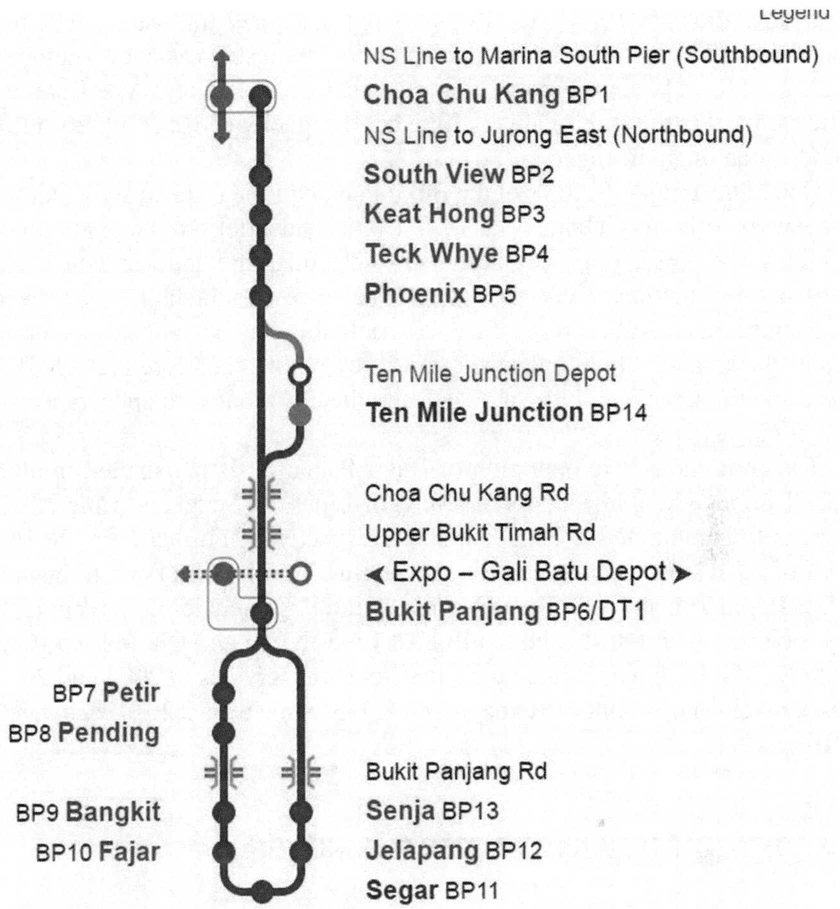

FIGURE 4.5   Bukit Panjang—Automated LRT in Singapore. *Source*: https://en.wikipedia. org/wiki/Bukit_Panjang_LRT_Line. Public domain.

Panjang ALRT line connects MRT Line at Choa Chu Kang station in the northern end and loop around the Bukit Panjang, a new town served by many local and express buses.

Bukit Panjang ALRT is operated along a double track loop line with 14 stations. The operating period spans from 5:00 a.m. to 1:00 a.m. The headway for rush hour is about 2–4 minutes and the rest of the day 6 minutes. There are 19 rubber-tired vehicles that transport 10,000 peak hour riders in 2015, which may translate to an annual ridership more than 3 million assuming an average peak period loading of 10% of daily traffic. The average speed is about 25 kmph and it takes about 28 minutes to traverse the entire line. The service operates in a sophisticated zonal system. For example, a Zone 1 train stops at stations 6, 10, 11, 12, 13 and depot; a Zone 2 train stops at station numbers 6, 7, 8, 9, 10, and depot; and a Zone 3 train stops at station numbers 1, 2, 3, 4, 5, and depot. Figure 4.6 is a typical Bukit Panjang ALRT station. The car used in the Bukit Panjang Line has a capacity of 105, with 20 sitting and 85 standing passengers.

One of the unique features of the Bukit Panjang Line is the misty windows. As the route is only about 6 m away from residential buildings at certain sections, the ALRT trains are designed to use misting windows, which will turn on misty automatically when it approaches residential buildings (Macau Transportation Infrastructure Office, 2012). In this way passengers cannot see inside the residential buildings whose privacy will be therefore protected. At the same time, the operation of ALRT and the city's outward appearance are not greatly affected.

The continuous safe operation of Bukit Panjang has marshalled in more ALRT applications in Singapore. Punggol Line and Sengkang Line ALRT were both inaugurated in 2003 and 2005, respectively. Furthermore, the success of ALRT has encouraged Singapore Bus Transit (SBT) and Singapore Mass Rapid Transit (SMRT), both transit operating agencies in Singapore, to try their hands in DLM. The North East Line (NEL) and Circle Line (CL), both DLMs, have inaugurated their first segment services in 2003 and 2009, respectively and continue to expand as of 2015 (Simens Inc., 2014; Schwandl, 2014).

4.3 DETROIT DOWNTOWN PEOPLE MOVER

Different from transit scale applications of automated transit technology, which usually have fixed route; fixed schedule; and charge a transit fare, automated people movers (APM) usually serve as a secondary function to support main activities. Examples include downtown, airport, and major activity center (MAC) circulations, or shuttle between two or a few interesting points that may or may not charge a fare. APM usually uses intermediate capacity vehicles, smaller than DLM trains and larger than personal rapid transit (PRT) podcars.

As a direct product of the Downtown People Mover (DPM) program executed by the US federal government in the 1970s, the Detroit People Mover is one of the three DPM applications in America and opened its service in 1987. It is a fully automated guideway transit system operating on an elevated guideway that is 2.9-mile loop. The Detroit People Mover connects 13 stations through the Central Business District (CBD) of Detroit. It uses two-car trains running on an elevated one-way loop (Panayotova, 2003). Two of these stations, Millender Center and Cobo Hall, are integrated into

(a)

(b)

FIGURE 4.6   (a, b) Petir Station Along Bukit Panjang ALRT Line. *Source*: Google Map, 2015.

buildings. The technology is steel wheels on standard-gauge steel-rails with linear induction motors instead of normal traction motors. The trains run every 3 to 5 minutes throughout the day between 6:30 a.m. and 2:00 a.m. on week days and shorter period on weekends (Detroit Transportation Corporation, 2015).

As shown in Figure 4.7, eight of the 13 DPM stations in Detroit are connected by pre-existing structures, over 9 million square feet of commercial and office space, such as the Renaissance Center that houses General Motors Corporation's headquarters. The DPM enables office workers, shoppers, and visitors to travel in the downtown area with great ease as it takes 15 minutes to traverse the entire loop (Sullivan et al., 2005).

The initial capital cost of the Detroit People Mover was $200 million (Panayotova, 2003) while it needs $10 million a year to operate. Only 15% of the operating expenses are covered by the 75 cent fare and 50% discount for senior and disabled riders. The city supplies the rest of the funds, $8.5 million a year. A special highlight of the Detroit People Mover system is the art installation on each station as demonstrated in Figure 4.8.

Data from the first year of Detroit People Mover operation showed a reliability of 98% with an average patronage of 11,000 per day (Dutta et al., 1991). The system was originally intended to be the downtown distributor for a rapid-transit system, which was not built. The ridership fluctuated along with the economic uncertainties in Detroit. However, the fact is that the Detroit Downtown People Mover is still chugging along in the Heart of "Motor City" of America after almost three decades without any major incidents.

4.4 AUTOMATED PEOPLE MOVERS IN LAS VEGAS

Different from the individual systems in other sections of this chapter, APM applications in Las Vegas are introduced here due to their concentrated geographic location along the Las Vegas Strip. It is also attempted to showcase the diversification of APM technologies, which brought along enormous challenges when individual applications are not coordinated or connected with each other and/or with the cityscape.

The Las Vegas Monorail (LVM) is located on the Las Vegas Strip, with a total length of 3.9 miles. Running along an elevated guideway with an average height of 30 feet, the Las Vegas Monorail connects Sahara Station in the north end of the Las Vegas Strip with MGM Grand Station in the south. The system supplier is Bombardier Transportation, and the alignment is based on an existing monorail between the MGM Grand and Bally's.

Opened in July 2004, the seven stations along the Las Vegas Monorail provide easy access to several world-class resorts, hotels, and the Las Vegas

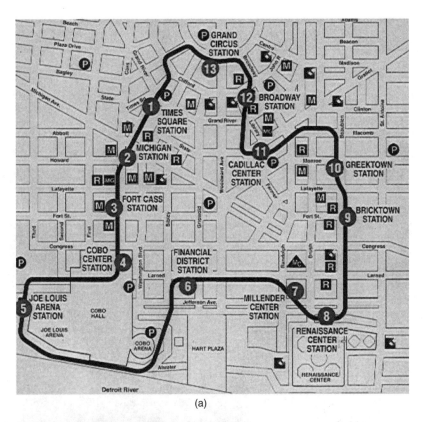

(a)

(b)

FIGURE 4.7   (a, b) Detroit Downtown People Mover (DPM).

(a) (b)

(c) (d)

FIGURE 4.8 (a–d) Art Work in Various Detroit Downtown People Mover Stations.

Convention Center. The basic alignment of LVM is shown in Figure 4.9. As the nation's first fully automated urban monorail transit system, the Las Vegas Monorail bears a price tag of $650 million and was completely funded by private entities (Snyder, 2005).

The LVM was originally envisioned as a joint venture between MGM Grand and Bally's Hotel, with the idea to create a one-mile transportation

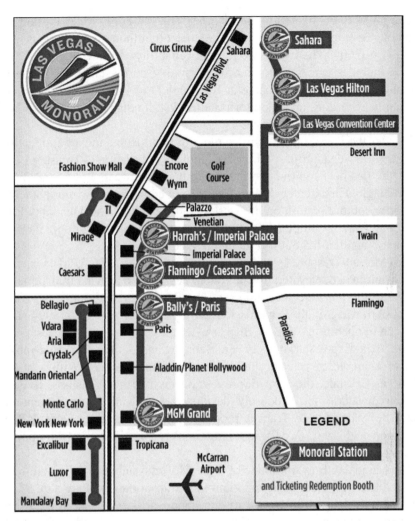

FIGURE 4.9 AGT Applications in Las Vegas. *Sources*: Vegas.com, 2009; Las Vegas Monorail Company, 2010. Public domain.

system linking the hotels but was expanded to have seven stations. The LVM utilizes 36 Bombardier Innovia 200 cars, nine trains with four cars each. Each climate-controlled train contains 72 seats, with standing room for an additional 152 riders. The single track guideway is about 30 feet in average above the ground and highest point of the guideway, Las Vegas Convention Center pedestrian walkway, is about 60 feet above ground. The driverless operation is enabled by an automatic train control (ATC) system. A suspension and guidance subsystem ensures stable lateral support. The all-electric, elevated train system has the smallest foot print, 26 inch-wide running surface (Las Vegas Monorail, 2015).

The LVM is contracted to Bombardier Transportation for operation and maintenance since 2009. It carried more than 60 million riders since its inauguration in 2004. The in-service rate, or reliability, is more than 99%. The system includes cameras on each train running during all hours of operation and closed-circuit TV cameras throughout the stations. Those cameras constantly recording and monitoring all ticket vending machines (TVMs), platforms, fare gates, and escalators.

Besides the LVM, there are three more tram or people mover applications along the Las Vegas strip. As shown in Figure 4.9, the Mirage–Treasure Island Tram shuttles between the two namesake hotels between 7 a.m. and 2 a.m. The Bellagio–City Center–Monte Carlo Tram connects a few more hotels in the north–south direction on the west side of the strip. The Mandalay Bay–Excalibur Tram is completely indoors, connecting the two main hotel resorts via Luxor, another large resort along the Las Vegas Strip.

The Mirage/Treasure Island Automated People Mover (APM) system is a fully-automated, elevated transit system designed to provide transportation along a single-lane, elevated guideway structure (Mori and Sandoval, 2011). The system was initially commissioned for passenger service in 1994 and it has been in operation for more than two decades.

The Aria Express tram between Bellagio and Monte Carlo casinos is another example of private automated transit applications. With a middle station at Crystals, the Aria Express connects two hotel/casinos, Bellagio and Monte Carlo via a densely developed retail and entertainment district. The Aria Express Tram is propelled by cable using two 460 KW ac motors on each train. Two 95-foot long trains run simultaneously in opposite directions.

The Mandalay Bay Automated People Mover system is a fully automated elevated transit system designed to provide transportation along a dual lane, elevated guideway structure (Mori and Sandoval, 2011). The system operates between the intersection of Tropicana and Las Vegas Boulevards and the Mandalay Bay Hotel/Casino. Having been initially commissioned for passenger service on April 2, 1999, the system has entered its second decade of operational life.

The piecemeal development of AGT in Las Vegas underlines a very important issue with automated transit or transit development in general—its coordination and interaction with other modes. As documented by previous studies (Liu, 1996; Liu et al., 1997), time loss and frustration associated with transfers between modes or even between vehicles within the same mode is a major impetus that discourages transit use. A true AGT can only establish its market when the transfer impetus is minimized, and travel time and reliability are superior or comparable to that of private automobiles.

With great variations of technology and disconnected alignment, the Las Vegas Monorail and tram trains may serve as a showcase of AGT applications at best. With very short route coverages, disconnected stations, and various ownership and operating schedules, the numerous AGT applications did not form a coherent transit network in Las Vegas, where the predominant modes are walking or automobile. It would be ideal if some coordination or integration was carried out during the development processes so that the AGT applications could form an integrated transit system with coordination and connection. Given the existing conditions, further integration and coordination maybe difficult but still possible, especially if public utility/welfare can be set up as a common goal.

4.5 DALLAS-FORT WORTH AIRPORT APM

The Dallas-Fort Worth International Airport (DFW), located between the cities of Dallas and Fort Worth, is the world's fifth busiest airport in terms of passenger enplanements. More than 65% of airport usage comes from connecting passengers who usually need to travel between terminals to make their transfers. In order to accommodate the needs of growing international travel demand and to limit transfer time to 30 minutes or less for en-route travelers, DFW chose an APM application, SkyLink, as an ideal connecting mode among all terminals (Corey, 2005). Figure 4.10 shows the SkyLink network layout.

At a price tag of $847 million, SkyLink, a fully automated people mover (APM) system, was constructed at the DFW International Airport and opened in the spring of 2005. The all-new elevated Skylink replaced the aging Airtrans, an early APM application, which had been in service since 1973 (Taliaferro, 2011). As one of the largest high-speed airport circulation systems in the world, SkyLink features a 4.7-mile double-loop bi-directional guideway that connects six terminals, including a future one, within 8 minutes. With a capacity of 69 passengers, average travel speed in the range of 35 to 37 mph, and headway of two minutes, SkyLink carries more than 3000 passengers each hour in each direction.

As a dedicated airside service, the Skylink APM serves passengers connecting between flights. There is no need for passengers to leave security and be re-screened when switching terminals. The system is only accessible airside and cannot be accessed by those not arriving at DFW or who have not been cleared security. Arriving international passengers, who are not pre-cleared, such as those from Canada, are security screened before access to their connecting terminals. Departing international passengers

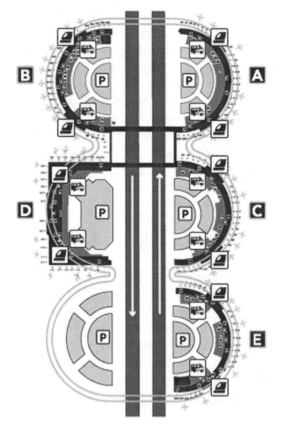

FIGURE 4.10 SkyLink in DFW International Airport. *Source*: Dallas-Fort Worth International Airport, 2008. Public domain.

connecting from domestic or pre-cleared international flights do not need to be re-screened, which provides expedited connections for the passenger flow.

4.6 AIRTRAIN AT JFK AIRPORT

The reason for the inclusion of the AirTrain at JFK International Airport in this chapter is its unique combination of airport APM and urban metro in one AGT technology application. The John F. Kennedy (JFK) International Airport, located in New York City, is the busiest international air passenger gateway to the United States (Federal Aviation Administration, 2015). The AirTrain at JFK International Airport is a fully automated guideway transit system that connects JFK to its adjacent cities. The 8.1-mile-long APM, which cost $1.9 billion to build, began construction in 1998 and eventually opened in December 2003. It has 10 stations with a 1.8-mile airport circulator loop and two extensions to urban transit systems that equal 6.3 miles.

The AirTrain uses AGT technology from Bombardier, and the capacity of the trains ranges from one to four cars with 75 to 78 passengers per car. The headway of the train is approximately 10 minutes, taking about 2 minutes between terminals. The AirTrain serves three main routes: All Terminals Route, Howard Beach Route, and Jamaica Station Route. As shown in Figure 4.11, the All Terminals Route is a circle route that connects all six terminal stations. The Howard Beach Route and the Jamaica Station Route connect the terminals and the regional mass-transit hubs, such as the New York urban subway and the Long Island Railroad (LIRR) stations.

The initial goal of AirTrain JFK was to provide air passengers and airport employees with rail access to JFK from the Howard Beach and Jamaica stations (Cerreno, 2009). When the Airport APM was linked to a broader vision for the redevelopment of Jamaica and Queens in New York, an automated transit application was born. As one of the nation's busiest transit hubs, Jamaica Station serves more than a quarter million commuters daily via three subway lines, 31 bus lines, and Long Island Railroad (LIRR) commuter rail lines. Shortly after the construction of AirTrain, the renovation of Jamaica station took place, which ushered in a series of infrastructure improvement projects, as well as economic development in the immediate surrounding areas. For example, the JFK Corporate Square, a 300,000 square feet office complex in downtown Jamaica was completed and it welcomed its first tenants in 2011 (Hendrick, 2011). In turn, further improvements to AirTrain JFK have been included in multi-million dollar joint capital projects by New York City and the Port Authority of New York and New Jersey (PANYNJ).

4.7 MORGANTOWN GROUP RAPID TRANSIT

The Morgantown People Mover or Morgantown Personal Rapid Transit System is an automated tri-mode (demand, schedule, and circulation) transit system. As explained in Chapter 1, the Morgantown People Mover should be classified as group rapid transit (GRT) because of its intermediate vehicle size and capacity. Each vehicle has a capacity of 21 passengers while 8 seated and the rest standing. Meanwhile, its operating characteristics are closely tied to Personal Rapid Transit (PRT) when it is on demand mode, that is, from origin to destination station directly.

The four-mile double track guideway connects five passenger stations and a maintenance facility. The guideway running surface is made of concrete, containing distribution piping for guideway heating to allow all-weather

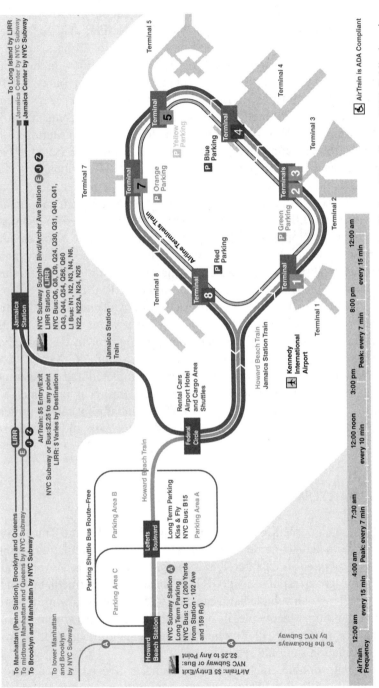

FIGURE 4.11 The AirTrain at JFK International Airport. *Source:* The Port Authority of New York and New Jersey, 2005. Public domain.

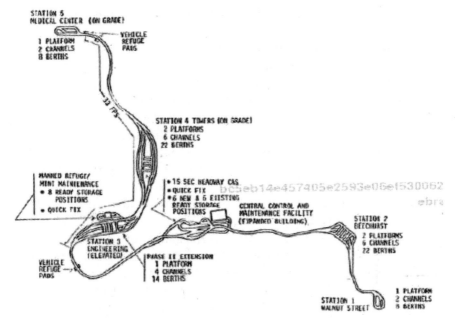

FIGURE 4.12   Morgantown GRT Corridor. *Source*: Yeschenko, 2013. Public domain.

operation. Inductive loops are installed inside the running surface to enable the transmission of messages between the vehicle and the control and communications equipment. Speed and station stop commands, as well as steering switch and calibration signals are received by the vehicle via inductive communication loops embedded in the guideway. Steering and electrical power rails are mounted along the side of the guideway.

Despite its aging and poor graphic quality, the sketch in Figure 4.12 maybe the best depiction of the schematic layout of Morgantown GRT with multi-berth, by-passing station layout. As shown in Figure 4.12, each station is composed of off-line channels for vehicle berthing, a key feature that affords direct operation between origin and destination stations.

The Morgantown GRT system consists of a fleet of 73 electrically powered, rubber-tired vehicles operating on a dedicated guideway under computer control (Schroder and Washington, 1980). It connects various activity centers of the university and community along the corridor. The present GRT operational system consists of 8.7 track miles of guideway, five stations, a vehicle maintenance facility with a small test loop and a central control facility. The Morgantown GRT vehicles logged approximately 1.5 million miles each year and has transported more than 67 million passengers without injury since its inauguration in the 1975 (Gannett, 2010).

4.8   ULTRA PRT AT HEATHROW INTERNATIONAL AIRPORT

As the origin of the modern automated transit concept, personal rapid transit (PRT), also referred to as automated transit networks (ATN), was conceived in the 1950s. The PRT concept has been in existence for more than a half of a century while a true PRT application eluded us. With its small vehicle, small footprint of guideway, and versatile operation characteristics, PRT promised the service qualities of private automobiles but the complexity of operations and lack of proven implementations in the real world made PRT application more like mirage than Promised Land for many transit agencies and passengers.

There are only three limited PRT applications around world as of 2016. They are located in Heathrow International Airport, London; Masdar City, Abu Dhabi; and Suncheon Bay, South Korea. All three applications are called PRT due to its small foot print of guideway and small vehicles with four- to six-passenger capacities. Each station at Heathrow has two alternative destinations that can be chosen. Both PRT applications in Suncheon Bay and Abu Dhabi have the capability to demonstrate station selection but it is not used by passengers due to the small or single line routing. All three of them are operating as single line shuttle services even though more ambitious or expansive networks have been planned in some locations, such as Masdar City. It is logical that PRT can be viewed as the natural evolution of APM—offering non-stop travel on demand—rather than something completely different or a competing technology.

On the other hand, it is argued that even though those applications to date are not seen as being very ambitious in term of showcasing the full ability of PRT to deliver high capacity, direct route, and flexible networks, they are the necessary first steps in proving the underlying principles of PRT (Pemberton, 2013). They are also valuable tools to help both operators and customers in building confidence in the technology and services. It is an incremental but necessary process to the Promised Land: ultimate application of full PRT systems. Therefore, Ultra at Heathrow International Airport in London is presented here.

As the main international airport in the United Kingdom and one of the busiest in the world, London Heathrow Airport was committed to the world's first PRT to provide key connectivity for the airport in 2005. As a pilot scheme, the initial application of Ultra PRT in Heathrow was designed to connect the Terminal 5 building with a commercial parking lot to explore the opportunities PRT may offer. The PRT service is designed to dramatically reduce the time that passengers need to move from their parked car to check-in counters.

Starting at a small testing scale, the initial Ultra PRT system has a 3.8-km (2.4-mile) double guideway that connects three stations, two in the T5

(a)

(b)

FIGURE 4.13   (a, b) Ultra PRT at Heathrow International Airport.

Business Car Park and one at Terminal 5 as shown in Figure 4.13. The Ultra PRT fleet is made up of 21 vehicles, and total travel time between the two terminals is about 5 minutes. The small footprint of PRT applications is well suited in this particular location because the current alignment traverses two rivers, seven roads, and green-belt land, not to mention negotiating aircraft flight paths and bridge in-ground services while conforming to the Terminal 5 architecture and appearance styles (Lowson, 2010).

The Ultra podcars are battery powered and can hold four adults and two children including luggage. Each podcar is controlled by an onboard computer with sensor systems. From the perspective of energy efficiency, it is supposed to save 70% of the energy compared with cars and 50% compared with traditional buses (Carnegie and Hoffman, 2007).

As the first commercially operational PRT system, Ultra at Heathrow International Airport transports 800 passengers per day between a Business Car Park and the Terminal 5. As observed by a passenger, the Ultra pods are electric and autonomous. The passenger tells the system where he or she wants to go using touch screens at the station. If a podcar is not waiting, one will arrive quickly—the average waiting time is less than 10 seconds. The control system arranges each "podcar" in relation to other pods before it sets off. However; once the "podcar" is underway, it is "thinking" for itself.

As of May 2013, the second anniversary of full operations, Ultra "podcars" have collectively operated more than 26,000 vehicle hours, transported more than 1.2 million passengers, and traveled more than 2.5 million Vehicle Kilometers, which is a milestone not only for Ultra but also for PRT applications in the world (Ultra Global Inc., 2015). Encouraged by the positive experiences from operators and passengers thus far, Heathrow Airport Holdings Limited is considering the expansion of the Ultra system (Ultra Global Inc., 2015).

REFERENCES

Brighinshaw, D. 2015. "Alstom wins 2 billion Euro Paris Metro train contract." *International Railway Journal*. January 30, 2015. Available at http://www.railjournal.com/index.php/rolling-stock/alstom-wins-%E2%82%AC2bn-paris-metro-train-contract.html. Accessed in May 2015.

Carnegie, J., and P. Hoffman. 2007. "Viability of Personal Rapid Transit in New Jersey." Prepared for New Jersey Department of Transportation by Alan M. Voohees Transportation Center, Rutgers, The State University of New Jersey.

Cerreno, A. 2009. "Integrated transportation and land use planning: facilitating coordination across and among jurisdictions." Project Final Report.

Corey, K. 2005. "DFW APM - Innovative solutions to success." In: Proceedings of Automated People Movers 2005. Moving to the Mainstream. 10th International Conference on Automated People Movers, Orlando, FL, 2005.

Detroit Transportation Corporation. 2015. "Schedule." Available at http://www. thepeoplemover.com/about-dpm/overview/. Accessed in October 2015.

Dutta, U., R. Tadi, and M. Keshawarz. 1991. "Detroit Downtown People Mover maintenance data: an overview." *Transportation Research Record*, vol. 1308, pp. 142–149.

Federal Aviation Administration. 2015. "Passenger boarding (Enplanement) and all-cargo data for U.S. airports." Available at http://www.faa.gov/airports/planning_ capacity/passenger_allcargo_stats/passenger/?year=all. Accessed in December 2015.

Gannett Fleming in Association with Lea + Elliott et al. 2010. PRT Facilities Master Plan.

Hendrick, D. 2011. "Jamaica's JFK Corporate Square gets first anchor tenant: MTA." *Queens Chronical*, July 11, 2011. Available at http://www.qchron.com/ digital_edition/. Accessed in June 2015.

Jonnes, J. 2010. *"Eiffel's Tower: The Thrilling Story Behind Paris's Beloved Monument and the Extraordinary World's Fair that Introduced It."* Viking.

Juillet. 2013. "Les compte des transports en 2012." Accessed in June 2015.

Las Vegas Monorail. 2015. "Operational experience." Available at http://www. lvmonorail.com/operations/. Accessed in December 2015.

Liu, R. 1996. *"Assessing Intermodal Transfer Disutilities."* University of South Florida.

Liu, R., R. Pendyala, and S. Polzin. 1997. "An assessment of intermodal transfer penalties using stated preference data." In: *Transportation Research Record 1607, The Journal of Transportation Research Board*, National Research Council, Washington, DC, pp. 74–80. Peer reviewed.

Lowson, M. 2010. "Preparing for PRT operations at Heathrow airport." In: Proceedings of TRB 89th Annual Meeting Compendium of Papers DVD, Washington, DC, 2010.

Macau Transportation Infrastructure Office. 2012. "Lessons from LRT development in Taiwan and Singapore." Accessed in March 2015.

Mancone, P. 2011. "Automation of line 1: a real opportunity for the Parisian Metro." In: Proceedings of APM-ATS Conference, 2012.

Metro-pole. 2011. "Metro accidents." Available at www.metro-pole.net. Accessed in August 2015.

Mori, J. D., and J. Sandoval. 2011. "Mandalay Bay People Mover: safe operations for 10+ years." In: Proceedings of the Thirteenth International Conference on Automated People Movers and Transit Systems, 2011.

Motty. 2015. "Paris metro ligne 14 - own work inspired by Image: Ligne 14.gif." Licensed under CC BY-SA 3.0 via Wikimedia Commons. Available at http:// commons.wikimedia.org/wiki/File:Paris_Metro_Ligne_14.svg#/media/File:Paris _Metro_Ligne_14.svg. Accessed in March 2015.

Panayotova, T. 2003. "People movers: systems and case studies." Facilities Planning and Construction, 2003.

Parkinson, T. E. 1986. "BC transit. the world's longest fully-automated metro." *Railway Gazette International*, vol. 142, no. 2. pp. 95–97.

Pemberton, M. 2013. "The track to Suncheon: making APMs intelligent." In: Proceedings of Automated People Movers and Transit Systems 2013: Half a century of automated Transit - past, present and future, Reston, VA. ASCE.

Robbins, A. S. 1996. "Singapore's first use of AGT in an urban environment." In: Proceedings of the 1996 Rapid Transit Conference, APTA, Atlanta, GA.

Schroder, R. J., and R. S. Washington. 1980. "Morgantown People Mover collision avoidance system design summary." Monograph. DOT. DOT-TSC-UMTA-80-37 Final Report.

Schwandl, R. 2014. "Singapore metro (MRT and LRT)." Available at http://www.urbanrail.net/as/sing/singapore.htm. Accessed in May 2015.

Shladover, S. 2012. "Literature review on recent international activity in cooperative vehicle-highway automation systems." Exploratory Advanced Research Program, Federal Highway Administration, U.S. Department of Transportation.

Simens Inc. 2014. "Simens in Singapore." Available at http://sg.siemens.com/aboutus/. Accessed in August 2015.

Snyder, T. 2005. "Las Vegas monorail innovations." In: Proceedings of Automated People Movers 2005. Moving to the Mainstream. 10th International Conference on Automated People Movers, Orlando, FL.

Sullivan, A., et al. 2005. "Detroit people mover: Automatic Train Control Upgrade (ATCU) Project." In: Proceedings of Automated People Movers 2005. Moving to the Mainstream. 10th International Conference on Automated People Movers, Orlando FL. 2005.

Swede, G. A. 1992. "Automated guideway transit systems. A compendium of AGT systems in operation, design and/or construction." Monograph. American Public Transportation Association.

SYSTRA. 2013. Paris Metro No. 14. MRT Technical Expert, June 2013.

Taliaferro, D. 2011. "DFW Skylink: enhancements to station quality." In: Proceedings of the thirteenth international Conference on Automated People Mover and Transit systems 2011: From People Movers to Fully Automated Urban Mass Transit, ASCE, May 22–25, 2011, Paris.

Trans.21. 2014. "Fall 2014 Airport APMs." Available at https://faculty.washington.edu/jbs/itrans/trans21.htm. Accessed in August 2015.

Ultra Global Inc. 2015. "Ultra provides environmentally sustainable 21st century transport solutions." Available at http://www.ultraglobalprt.com/about-us/. Accessed in March 2015.

Yeschenko, B. 2013. "Train control upgrade for the Morgantown personal rapid transit systems at West Virginia University." In: Proceedings of Automated People Movers and Transit Systems 2013: Half a Century of Automated Transit – Past, Present, and Future, Reston, VA.

CHAPTER 5

CHARACTERISTICS OF AUTOMATED TRANSIT APPLICATIONS

Building on the particular examples of various automated transit applications documented in Chapter 4, this chapter will present a general landscape of automated transit applications in terms of system, service, and financial characteristics. With the wide span of vehicle and network capacities ranging from very small podcars to very long driverless metro (DLM) trains; varied climate, culture, and institutional structures; and diversified operating characteristics, the automated transit industry has the potential to provide a wide spectrum of applications for different government, institutional, and even private entities to choose from.

5.1 SYSTEM CHARACTERISTICS

Most layperson can easily distinguish between the various automated transit applications involving large and/or small vehicles. Equally important elements of an automated transit application are the network configuration, fleet size, station layout, and control and communication components. Each of these may be a key element that determines and distinguishes each automated transit system from another. It is easy to assume that large vehicles transport a large number of passengers and vice versa. However; the true capacity of a given automated transit system is actually dictated by the throughput, which

Automated Transit: Planning, Operations, and Applications, First Edition. Rongfang (Rachel) Liu.
Copyright © 2017 by The Institute of Electrical and Electronic Engineers, Inc. Published 2017 by John Wiley & Sons, Inc.

is determined by not only the size of vehicles but also the frequencies or headways between the vehicles passing a selected point on the line. Station throughput capacities and network configurations are equally important to determining the overall system throughput "capacity." Detailed calculations are presented later in the chapter to demonstrate the process or key elements that can be adjusted to increase the throughput of an automated transit application.

5.1.1 Physical Layouts

A quick scan of existing automated transit literature often reveals that the physical layouts of automated transit systems are classified into shuttle, loop or circulator, line-haul, and network, which are not only manifested via different shapes and sizes of the track or guideway layout but also associated with different operating characteristics. As demonstrated in Figure 5.1a, a simple single-lane shuttle system has two terminal stations and one vehicle shuttling back and forth between those two stations. With only one guideway track, the simple shuttle system is limited to one vehicle that can be operated at any given time. This configuration may have limited capacity or excessive empty runs if the passenger flow is not bi-directionally balanced.

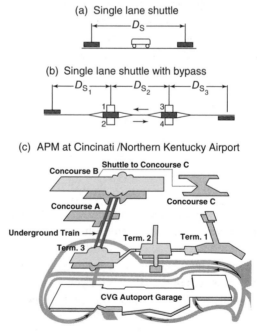

FIGURE 5.1 Various Single-Lane Shuttle Configurations. *Source*: Anderson, 1978 and Lea + Elliott, 2010.

An economical expansion of the simple single-lane shuttle system can be accomplished via bypasses at intermediate stations, which results in a simple single-lane shuttle with Bypass, as shown in Figure 5.1b. The added bypasses at intermediate stations allow more vehicles to operate simultaneously without adding the second full-length guideway track. The airside automated people mover (APM) at Cincinnati/ Northern Kentucky International Airport (CVG) is an example of dual-lane shuttle system, which is made of two single lane simple shuttle operations, as shown in Figure 5.1c.

Another common automated transit guideway configuration is the loop system with single or dual lanes. Minor variants to the loop configurations are shown in Figure 5.2a. The key difference between a shuttle and a loop system is the directionality of vehicle travel along the guideway lanes. In a shuttle system, all vehicles travel both directions along the majority of each guideway lane, whereas in loop systems all vehicles travel in one direction only along all or a majority of the guideway lanes. A dual-lane loop is usually made of two circular tracks, inner and outer, so vehicles circulate to all the same or slightly different station locations via opposite directions between the two tracks. The Skylink in Dallas-Fort Worth International Airport (DFW) is a good depiction of dual-lane loop system, as shown in Figure 5.2b.

A further variation from the single-lane loop may produce a line-haul configuration depending on the operation at the terminal stations. As demonstrated in Figure 5.3a, a line-haul configuration is formed by collapsing the single-lane loop into a corridor with side or central platforms. The line-haul configuration is often seen in urban transit systems, such as Honolulu Automated Rail Transit in Honolulu Hi, and Paris Metro Line No. 1 and No. 14.

Furthermore, connectivity and transfer ability are important characteristics that distinguish urban transit systems from simple shuttle or loop people movers. When more than one line-haul transit route, that is, trains operating on a fixed-route configuration, are interconnected via transfer stations, a transit network system is formed. Figure 5.4b shows the Copenhagen Metro Network, where Lines M1 and M2 are already completed DLMs and Lines M3 and M4 are still under construction as of 2016.

As illustrated in Figure 5.4a, an idealized transit network can be created to serve various sizes and shapes of urban developments and major activity centers in real world applications, which are not confined to linear corridors. With the advent of automated transit technologies that operate in a demand–response dispatch mode, new types of operations that deviate from a fixed-route operation are being considered. An automated transit network is created when more than one automated or driverless transit routes are created to converge and connect, or when vehicle routing is dynamically changed in a demand–response dispatch mode. There are few automated transit applications configured as networks in existence as of 2016, with most existing

(a) Loop configurations

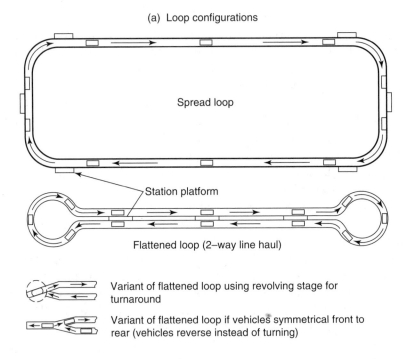

Spread loop

Station platform

Flattened loop (2–way line haul)

Variant of flattened loop using revolving stage for turnaround

Variant of flattened loop if vehicles symmetrical front to rear (vehicles reverse instead of turning)

(b) Example of loop configurations

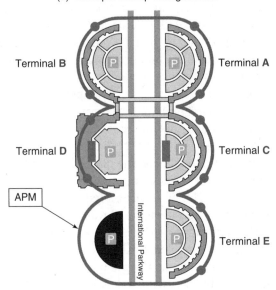

FIGURE 5.2   Examples of Loop System. *Source*: Irving et al., 1978 and Lea + Elliott, 2010.

(a) Line haul configurations

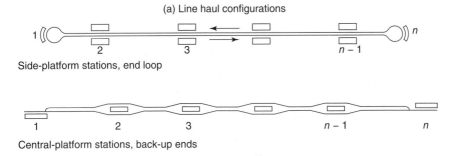

Side-platform stations, end loop

Central-platform stations, back-up ends

(b) Example of line haul configuration

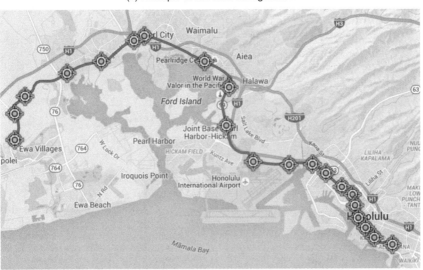

FIGURE 5.3 Line-Haul Configuration Scheme and Example. *Source*: Anderson, 1978 and Honolulu Authority for Rapid Transportation, 2016.

automated transit systems currently operating in linear corridor applications. But the number of networked automated transit systems will grow as more transit agencies convert or build more automated transit services within metropolitan areas and begin to connect them together as in Copenhagen.

When offline stations are built and utilized in combination with demand–response dispatching of automated transit vehicle/trains, promises of direct origin to destination travel within automated transit network applications can be realized. This concept of "personalized" transit service was conceived several decades ago and given the designation of PRT. Regrettably, there are still no large-scale PRT networks like that illustrated in Figure 5.4a as of 2016. The Morgantown group rapid transit (GRT) did build and occasionally uses

(a) Network configuration

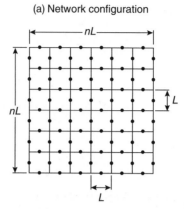

(b) Example of automated transit network: Copenhagen metro

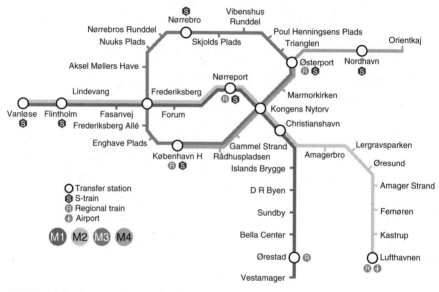

FIGURE 5.4 Automated Transit Network Scheme and Example. *Source*: Anderson, 1978 and Lea + Elliott, 2010.

its offline stations, but the single-line configuration with five stations rarely provides the opportunity for testing bypassing, connecting, or transferring capabilities for lack of network coverage.

5.1.2 Scale of Systems

With the recent expansion of DLMs around the world, especially in Europe, Asia, and South America, the DLM track miles and number of stations, as well as vehicle fleet are expanding rapidly. According to the International

Union of Public Transport (UITP, 2013), the total route or track kilometers for DLM grew exponentially from 1980 to the early 2010s. Individual route length ranges from 2.2 miles in Hong Kong Disney Resort Line to 33 mile for the Dubai Red Line.

Compared to the rapid expansion of DLM in urban areas, airport APM is relatively stable and still dominates the number of applications within automated transit categories. Due to the various sizes, enplaned volumes of passengers, and diversified functions of APM systems, the scale of the APM in an airport also spans a wide range as demonstrated in Figure 5.5a. The number of stations ranges from 2 to 16 and the length of each systems spans from 0.2 to 8.1 miles. Among all the existing APM systems in North America, the AirTrain at John F. Kennedy International Airport (JFK) has the

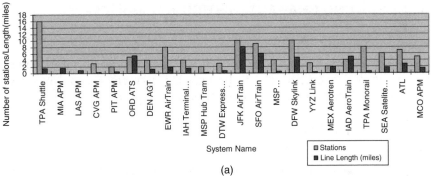

(a)

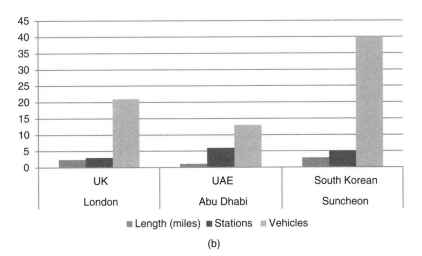

(b)

FIGURE 5.5 (a, b) Scales of DLM and Airport APM Applications. *Source*: Gambla and Liu, 2012.

longest alignment length, 8.1 miles, with three service routes, and 10 stations. It is noteworthy that this largest airport APM is also the smallest DLM categorized by UITP as a metro system, since it travels several miles off the JFK Airport property to connect with a major commuter rail station in Queens, New York.

Contrasting to the exponential growth of DLM and steady expansion of APM applications, PRT applications are limited both in terms of the number of applications established and the scale of projects being implemented. As documented in the last chapter, the Ultra PRT at Heathrow International Airport in London represents the typical scale of PRT applications implemented thus far: track length ranges from 1.1 to 2.9 miles with three to six stations and 12 to 40 podcars, as exhibited in Figure 5.5b.

5.2 OPERATING CHARACTERISTICS

Once the physical layout and scales are set, an automated transit application may adjust its capacity and service quality using different operating strategies. A few key concepts need to be introduced before general operating characteristics of automated transit applications are presented:

- Operational headway—the time between consecutive trains on a common route.
- Platform headway—for synchronized double shuttles only, the time between successive trains departures from a station platform without regard to the platform side from which the trains depart.
- Station dwell time—for each station, dwell time is the time during which the train is stopped at the station, usually measured in seconds.
- Travel time—the time between sequential stations, usually measured in minutes from the point in time a train's doors are fully closed and begins its movement, to the time it breaks to a full stop and begins to open its doors at the next station.
- Wait time—for passengers, the time spent waiting on the station platform until a train arrives that they can board that will take them to their destination station.
- Trip time—for passengers traveling on a single line, the sum of all station-to-station travel times plus the dwell times at all stations; for passengers transferring between lines to connect pairs of origins and destinations, the total time to make the trip, including both trip time on each line and the wait time at the connecting station.
- Round trip time—for each route, the time it takes a train to complete one circuit.

- Line capacity—the number of passengers per hour that can be carried past a given point on each independent route, usually measured in persons per hour per lane per direction.

5.2.1 Operating Strategies

As observed in the real-world applications, there are basically three types of operating strategies for automated transit applications: continuous, scheduled, and on-demand (Davis, Love, and Sturgell, 2013), which are defined as the following:

- **Continuous service.** Continuous service is when trains operate continuously throughout the system at regular intervals without stopping for extended periods of time and regardless of whether passengers are riding in or waiting for trains. Trains are separated by regular time intervals known as headways.
- **Scheduled service.** Scheduled service is when trains operate on a given or predetermined schedule and stop once a route has been completed, or between route operations, regardless of whether passengers are riding in or waiting for trains. Once trains depart station platforms they travel along predetermined routes stopping at station platforms on their route to allow the transfer of passengers. When trains arrive at their final destination, they remain parked until the next scheduled departure.
- **On-demand service.** On-demand service is when trains operate on a route only if there is a service request by a passenger. This can be used for simple shuttle systems, or for networks where trip "demands" served are destination-specific, in that trains only travel to requested stations, departing from stations where travel requests were made, or along predetermined routes stopping at all stations along the route. Once trips have been completed, trains stop until another travel request is activated.

As part of the urban transit system or network, all of the DLMs operate on fixed route, fixed schedule. Most APM systems, especially those serving airports, institutions, or major activity centers (MAC), use continuous modes with fixed or variable headways. The Morgantown GRT is the only system that uses all three modes of operation: demand mode, schedule mode, and continuous circulation mode (Raney and Young, 2005).

In the case of Morgantown GRT, demand and schedule modes are designed to operate during periods of peak demand. Continuous circulation mode is used during off-peak service. The demand mode attempts to capture the on-demand aspect of PRT/GRT. The other two modes are prescheduled vehicle operation patterns intended either to optimize throughput during peak demand or to limit operating expenses during off-peak hours.

The following practices are observed in Morgantown GRT but can be applied to other systems if corresponding modes are chosen. Demand mode reacts dynamically to passengers' request for service. The algorithm governing the demand mode balances two parameters—passenger wait-time and vehicle occupancy. Once a passenger enters a station and requests service to a destination, a timer starts. If the timer reaches a pre-determined limit, typically 5 minutes, a vehicle is activated to service the request even if no other passengers have requested the same destination. Also, if the number of passengers waiting to travel to the same destination exceeds a pre-determined limit, usually 15 passengers, a vehicle is activated. Once activated, a vehicle opens its doors and an electronic display prompts passengers to board. The vehicle doors remain open for 20 seconds allowing passengers to board. The doors close automatically, and the vehicle departs to its final destination, avoiding any intermediate stations. The two parameters that govern the algorithm, maximum passenger wait time and vehicle occupancy, can be varied by central control. On cold winter days maximum passenger wait is reduced in order to minimize discomfort to passengers while waiting at the stations.

When schedule mode is applied in Morgantown GRT, vehicles travel directly from origin to destination based on predetermined schedules. For high demand periods with well-known travel-demand patterns, schedule mode operates slightly more efficiently than demand mode. During peak demand periods, operating in either demand or scheduled mode, the system transports approximately 1500 passengers per hour. Historically 80% of travel demand is between the Beechurst and Towers Stations (Raney and Young, 2005). As Morgantown GRT primarily serves students from the University of West Virginia, periods of peak demand coincide with the 20-minute period between scheduled classes. The average waiting time for passengers traveling between the Beechurst and Towers Stations during peak demand is about 1 minute. If the system is in demand mode, the 15-person rule is usually triggered after about 60 seconds. During the 20 seconds in which the vehicle doors are open, more passengers may arrive and board, often filling the vehicle to capacity. Wait time is higher at less busy stations.

During off-peak hours, demand mode would result in many nearly empty vehicles traveling about the system. During periods of low demand, the Morgantown GRT switches to continuous circulation mode which operates like a local bus, stopping at each station along the route on a preset schedule. Passenger travel time to destination increases while the system operates more cost-effectively.

Similar to the on-demand operation of GRT applications, PRT vehicle responds to individual demand whenever it is requested. Given the small vehicle size and personal nature of PRT applications, the PRT vehicle usually respond immediately. Without any restriction from occupancy or waiting time threshold, the real-world operation of PRT in Heathrow International Airport resulted in much less waiting, usually less than 30 seconds for passengers. The prompt response/departure, when combined with nonstop travel and seated rides, results in a level of service far higher than any other systems.

According to Andreau and Ricart (2010), the operating efficiency of metro or subway systems is closely linked to the availability and scheduling of drivers. The DLM has the potential to improve efficiency by eliminating the constraint of driver availability. Several studies (Graham et al., 2009; Litman, 2008; Paulley et al., 2006) show that increasing train frequency improves service quality by reducing waiting time, which may have the potential to increase travel demand. For example, operating trains of half the length at twice the frequency should usually increase ridership without increasing traction energy cost. The barrier to operate trains at higher frequencies on conventional metro lines is that driver costs will double assuming more drivers are available. For DLM operations, there is no additional driver cost, which may provide great incentive to employ the high frequency operating strategy in order to generate more demand and improve quality of service.

5.2.2 Station Operations

As explained earlier, one of the station design characteristics, off-line station, separates PRT from existing DLM or APM applications. There are a few examples of off-line station design, which include Morgantown GRT and all three PRT applications, but Morgantown GRT remains the only one that implemented on-demand mode, which utilizes off-line station/berth design and is described here.

The three middle stations of the Morgantown GRT are of relatively complex design. Figure 5.6 shows the Towers Station, which was constructed to meet significantly higher demand to the west than to the east. The Towers Station has six switch points, S1–S6, six merge points, M1–M6, and six channels, Ch1–Ch6. Each channel or docking area has three or four vehicle berths. Switch points provide opportunities for a vehicle to either bypass a station or to dock at a channel.

For example, a vehicle traveling from the Engineering Station to the Medical Station bypasses the Towers Station by taking the outside fork of S1. After bypassing the Tower Station, the vehicle will merge with entering traffic at merge point M2. A vehicle entering the Tower Station en route from the Engineering Station may stop at any of channels 1 through 4. After passengers unload and new passengers board, channels 1 through 3 divert the vehicle back toward the Engineering Station while channel 4 allows the vehicle to continue to the Medical Station. At each vehicle channel, all but the first berth are reserved for unloading. The first berth in each channel is used for loading

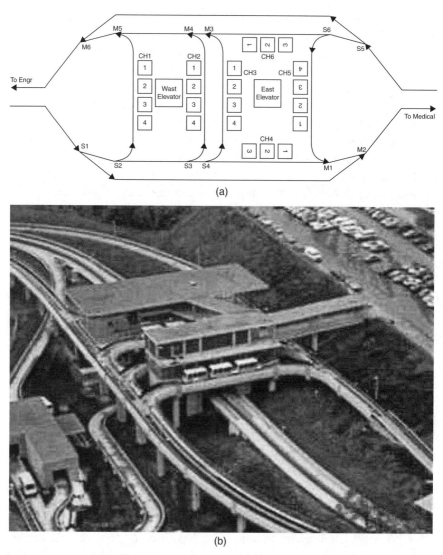

(a)

(b)

FIGURE 5.6 (a, b) Middle Station Design and Photo, Morgantown GRT. *Source*: Raney and Young, 2005. Public domain.

passengers. Vehicles have automated doors on both sides. The configuration of the station and channel determine the set of doors that is activated. For the Towers Station shown in Figure 5.6, the doors open to the left in channel 1, and to the right in channel 2.

All stations are constructed so that pedestrians are prohibited from crossing the guideway. As seen in Figure 5.6a, channels 1 and 2 share a passenger platform that can be accessed from underneath via stairs and the West Elevator.

Channels 3 through 6 share a passenger platform that can be accessed via stairs and the East Elevator.

The physical dimensions of the intermediate stations along Morgantown GRT are roughly $200' \times 120'$ including platforms and channel guideway. The size is dictated by the need to accommodate the various channel movement and expected capacity. These dimensions do not include deceleration/acceleration lanes of the guideway. Figure 5.6a is not drawn to scale. The segments from S1 to S2 and S5 to S6 are the deceleration ramps. A vehicle traveling at 30 mph requires 500 feet, at a smooth deceleration rate of 2 feet per second/second (fps^2), from S1 to S2 to slow to the channel entrance speed of 5 mph. The acceleration ramps, M1 to M2 and M5 to M6, need to be comparable in length since acceleration to cruising speed is also 2 fps^2.

The Engineering and Beechurst Stations are similar in design to the Towers Station. The layout differences reflect the differences in the expected volume and distribution of trips. Also, bypass lanes at the Engineering and Beechurst Stations are routed underneath the station as shown in Figure 5.6b. The Engineering Station also encompasses one of the two maintenance facilities.

5.2.3 System Capacity

Similar to all transit operations, the capacity of any automated transit system is a function of vehicle size and service headway intervals between units (Grava, 2002). Different from traditional manually driven transit, automated transit applications have potential to operate much shorter headways due to automated central controlled vehicle operations. Furthermore, the offline station design for PRT systems has the potential to bypass any intermediate stations and transport passengers from origin directly to destination, which remains as theoretical in lieu of real world applications.

For a typical line-haul configuration, the passenger throughput per hour per direction may be derived via the following equations:

$$C_P = C_V \times F_V \qquad (5.1)$$
$$F_V = T/H_V \qquad (5.2)$$

where:

C_P: System throughput, passengers per hour per direction

C_V: Vehicle capacity, passengers per vehicle (train)

F_V: Vehicle frequency: vehicles (trains) per hour

T: Minutes or seconds per hour, 60 minutes or 3600 seconds per hour

H_V: Vehicle (train) average headways in minutes or seconds

For example, a six-car train with 100 passenger spaces in each car, running at 1 minute headway could achieve a system capacity of 36,000 passengers per hour per direction as derived here:

$$\text{Vehicle capacity, } C_V = 6 \times 100 = 600,$$
$$\text{Vehicle headway, } H_V = 1 \text{ minute;}$$

therefore,

$$\text{Vehicle frequency, } F_V = 60 \text{ trains per hour and}$$
$$\text{Time per hour, } T = 60 \text{ minutes}$$
$$C_P = C_v \times F_v = C_v \times T/H_V = 600 \times 60/1$$
$$= 36{,}000 \text{ passengers per hour per direction (pphpd)}$$

To ensure the safety spacing between two successive vehicles traveling on the same alignment, a general equation is used to calculate the theoretical minimum allowable headway:

$$H_V = T_D + V_I/2a_F + D_S/V_I + L/V_I + V_I/2a_I \tag{5.3}$$

where:

H_v: Vehicle headway

T_d: Communication time delay

V_I: Initial velocity of the vehicle, assuming to be equal for all vehicles

a_f: Deceleration rate of the following vehicle

D_s: Safety distance behind rear bumper of lead vehicle

L: Length of vehicle, assuming to be equal for all vehicles

a_l: Deceleration rate of the lead vehicle

As shown in Table 5.1, it is theoretically possible to derive a minimum allowable headway of 3 to 25 seconds for PRT and DLM, respectively, when plugging in all the parameters into the above equation. However, practicality and safety precautions, which allow for multiple performance degradation and failure modes, have precluded any implementation of headways in the single-digit second range. In reality, most line-haul transit applications, operating entrained vehicles with two to six cars have minimum design headways of 60 to 90 seconds. These larger systems typically have peak hour headways of 3–5 minutes, and are capable of transporting more than 10,000 passengers per hour per direction.

TABLE 5.1 Theoretical Minimum Headway for Automated Transit Applications*

Types of Applications	DLM	APM	PRT
Communication Time Delay (T_D, second)	2	2	2
Initial Velocity of Vehicles (V_I, ft/s)	44	29	22
Deceleration rate of the following vehicle (a_F, ft/s^2)	1	2	10
Safety distance (D_S, feet)	3	2	1
Length of vehicle (L, feet)	240	120	20
Deceleration rate of the lead vehicle (a_L, ft/s^2)	4	9	15
Theoretical minimum allowable headway (H_V, second)	25	12	3

*Note: The calculation represents minimum theoretical headways achievable when failure modes are not taken into full consideration.

APM shuttles or circulators, which usually use shorter trains with less vehicle capacity but shorter headways, especially during the peak hour operations, are able to achieve similar system throughput as shown in Table 5.2. The system throughput for PRT applications is significantly smaller due to its smaller vehicle and practical headways. However, while it is understood that the advantages of PRT system is not throughput but on-demand or direct origin to destination (O-D) travel when a comprehensive network is developed, the practical capacity of individual high demand stations is typically the most important metric to consider.

Comparing existing APM and PRT applications, it is observed that there are significant overlaps between APM and PRT in terms of line capacity range. For example, APM capacity ranges from 2000 pphpd for small monorail applications, to 2500 pphpd for cable-propelled systems, to 6000 pphpd by steel wheel applications. PRT capacity ranges from 2400 pphpd for Ultra in Heathrow International Airport to 5400 pphpd for Vectus in Suncheon Bay (Muller, 2015).

TABLE 5.2 Typical Peak Hour Capacity for Various AT Applications

AT Category	Number of Cars Each Train	Fleet Configuration	Vehicle Capacity	Headway (minute)	Peak Hour Capacity (pphpd)
DLM	6	Entrained vehicles	700	3	14,000
APM	2	Entrained vehicles	200	1	12,000
PRT	1	Single vehicle	6	0.5	720

5.3 FINANCIAL CHARACTERISTICS

Aside from the hard-to-reach objective of direct origin to destination travel as conceived for PRT systems on a very large scale, high cost is another criticism of automated transit applications, especially AGT systems implemented in the United States. It is agreed that quite a few AGT applications, including Morgantown GRT and Metromover in Miami, have experienced major capital cost overruns (Office of Technology Assessment, 1975). However, different scholars, public agencies, and manufacturers interpret the cost data differently, therefore pinpoint the reasons for cost overrunning differently. Based on previous studies and system data available via government sources and author's research, this section attempts shed some light on the capital, operation and maintenance, and life span cost of AGT applications. These reinterpreted costs may establish the basis for comparison and evaluation between automated transit technologies and conventional transit applications, and between operating modes and system configurations in the following chapters.

5.3.1 Capital Investment

Similar to conventional rail transit investment, the capital costs for automated transit applications are made up of initial purchase and installation of vehicles, guideways, tracks, power, stations, signals, and communications systems. It also contains the cost for purchasing right-of-way, utility relocation, environmental mitigation, as well as design, financing and administrative expenses associated with organizations and processes that execute the automated transit implementation procedures. The "soft costs" may be proportional to the scale and duration of the project design and implementation, but it may not always be the case.

The general approach for estimating capital costs for automated transit applications is to develop detailed costs for a specific system design-point, which is referred to as the baseline design, and to define the variation in system component cost with production quantity (Irving et al., 1978). These analyses are synthesized to produce a mathematical model with the capability to predict system costs in terms of aggregation effects.

There is no official systematic track of capital investment on automated transit applications due to its various public and private funding sources, and the special development and demonstration stages of advanced technology projects. Only scattered estimates on a few selected applications are gleaned through various project reports and research papers, which are presented here to demonstrate the magnitude and components of capital investment for automated transit applications.

Using Vancouver Automated Transit Line, one of the earliest and longest fully automated metros in the world, Parkinson (1986) documented that the as-build cost for the 50-km track system with 15 stations is around 854 million Canadian dollars in the late 1980s. As shown in Figure 5.7, the largest capital

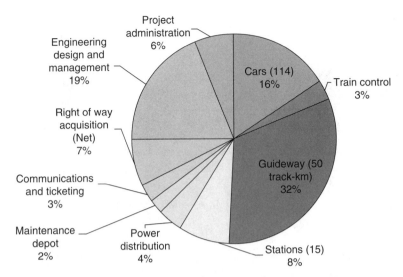

FIGURE 5.7   Capital Investment Components and Magnitudes. Data from Parkinson, 1986.

cost category is from the guideway, about one-third of the total capital cost or 40% when right-of-way acquisition cost is included. The second largest category is engineering design and management, almost one-fifth of the total capital investment or about one quarter when the administration is combined with the design and management. The next big ticket item, vehicle cost, should not be any surprise to count about 16% of the total capital investment for the Vancouver system.

The basic categories of capital investment for different applications are identical and proportions may be similar, even when the magnitude of total costs may vary dramatically. For example, the first phase of the Metromover in Miami, with 1.9-mile two-way loop and nine stations, cost $159 million to complete in 1984 (Holle, 2015). The two Miami Metromover extensions: the Omni leg with six stations and 1.4 miles of guideway and the Brickell leg with six stations and 1.1-mile guideway cost a total of $248 million in 1994. Allocating the total capital investment to unit length of guideway miles, the unit costs for the initial segment and extension are $84 and $99 million per mile, respectively. Besides the apparent configuration differences, both phases have implemented exactly same mode of operations, same guideway technology, and same type of vehicle. Therefore, it is safe to suggest that the time differences and associated manufacturing and construction cost escalation played an important role in causing the unit price differences.

An early APM case study (Shen and Huang 1995) has gathered capital cost data for Vancouver Skytrain, Lille VAL, and Taipei APM applications. As shown in Figure 5.8, the total costs for those applications range from $920

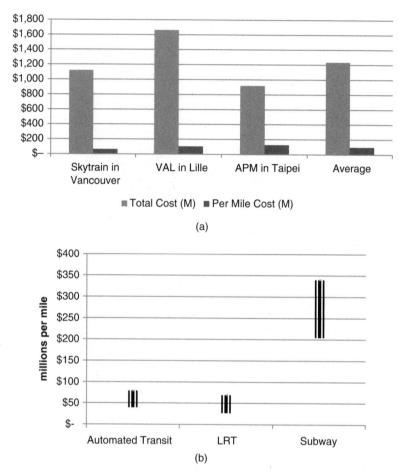

FIGURE 5.8 (a, b) Capital Investment Ranges. Data from Shen and Huang, 1995.

million to $1.66 billion in 1994 US dollars. The average unit costs range from $60 million to $125 million per track mile. When compared with other conventional rail transit, such as light rail or subway, the capital investment per track mile for AGT is very similar to that of LRT and much cheaper than heavy rail or subway as shown in Figure 5.8b.

A recent study of DLM applications provided estimates of incremental differences between conventional metro and fully automated or DLMs in capital investment (Cohen et al., 2015). As shown in Figure 5.9, many automated transit applications in Asia, Europe, and America metro-scale systems may cost the same as its conventional manually driven metros. Some automated

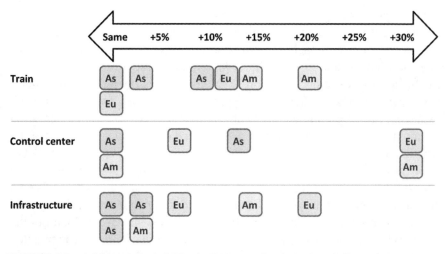

FIGURE 5.9   Additional Capital Cost for DLM Applications. *Source*: Cohen et al., 2014. Reproduced with permission of Transportation Research Board.

transit applications, especially those located in Europe and America, may cost 30% more than their traditional transit counterparts in the same locations.

Despite the much hyped criticism of high costs for automated transit applications in the United States, not many serious studies examined or tried to understand the reasons that caused the high costs or cost over runs in the first place. Looking back, it is noted that the over design of Morgantown GRT system may have doubled the cost of the guideway and vehicles (Anderson, 1996). A good portion of the Morgantown GRT capital investment maybe avoided if it was not mandated to meet the Nixon Inauguration date. The expedited process certainly has increased the cost. A follow-up study (Hsiung and Stearns, 1980) estimated that approximately one-third of the capital investment for Morgantown GRT could be partially attributed to research and development costs.

5.3.2 Operating Expenses

Operation and maintenance (O&M) costs are important for transit projects since they can constitute as much as 80% of annual expenditures to operate a service (Meyer and Miller, 2001). Comparing to the conventional transit operations, automated transit applications should incur less operation expenses since drivers are no longer needed. On the other hand, the number of personnel needed for central control and dispatch under fully automated operations may require fewer people, but with higher qualifications and correspondingly higher wages. Therefore, the balance of operation and maintenance expenses for each application can vary significantly and should be examined closely.

Two common approaches for preparing O&M cost estimates are cost-allocation models and resource build-up models. Cost allocation models are usually used in existing operations, where previous operation and maintenance expenditures are allocated to appropriate categories. On the other hand, resource build-up models are applicable to new services so the total operation and maintenance costs are estimated based on the individual line items of the overall operations.

Operating costs estimated for automated transit application were built up in terms of fixed and variable cost categories. Costs of all labor, materials, and electrical power not related to vehicle fleet size are included in the fixed costs, and all operating costs which are a function of fleet size are included in the variable costs. Table 5.3 summarizes the O&M costs that have been aggregated in a specific automated transit application during the 1970s. For example, the guideway O&M costs include automatic spray painting, internal cleaning of magnets and conductors, and checkout and re-magnetization of the ferrite magnets. Station maintenance and operating costs include cleaning, electrical power, fare collection equipment, elevators, lights, and air-conditioning, and maintenance of all non-electronic equipment. The electronics operating costs

TABLE 5.3   Operating Cost Elements

Fixed Operating Costs (labor, materials and power not related to vehicle fleet size)
        Guideways
                Painting
                Cleaning and maintenance
        Station maintenance and power
                Elevators/lighting/doors/thermal control
        Electronics (operation and maintenance)
                Computers
                Fare equipment
                Control sensors and instrumentation

Variable operating costs (labor, materials and power related to vehicle fleet size)
        Vehicles
                Power
                Maintenance
                Cleaning
        Vehicle cleaning and storage facilities
        Vehicle maintenance and overhaul facilities

General administration and overhead costs (administration personnel, insurance
    expenses, employee benefits, and other administrative expenses)

Source: Irving et al., 1978.

cover inspection and upkeep of station fare collection equipment, computers, communications links and buffers, and guideway-mounted sensors and instrumentation.

Another way to aggregate the O&M cost of automated transit applications is to categorize them into four different groups (Federal Transit Administration, 2010):

- Labor costs
- Material costs
- Utility costs
- Administration costs

The first category is labor costs, which include all the wages paid to workers who are involved with daily automated transit services. The labor costs, both in direct wages and indirect benefits, are extremely high, and may make up as much as 85% of the total operating and maintenance costs of a system. Based on the results of a recent airport APM study (Gambla and Liu, 2012), labor cost constitutes about 75% of annual airport APM O&M cost. This cost can come from various sources, such as salary for maintenance crew and management staff. The maintenance crew might include technicians, electricians, mechanics, operation engineers, and controllers, etc. The management staff was limited to personnel actively controlling and directing an APM system operation.

The material costs cover all the materials consumed during maintenance, operation, or inspection processes. The third category is utility costs, which is similar to conventional mass transit systems, and covers payments made to utility companies for the utilization of their resources such as electric, gas, water, telephone, and other communication needs (Federal Transit Administration, 2010). It is noted that the cost of propulsion power, electricity, comprises the largest portion of this cost. The last cost category, administration cost, contains all the other miscellaneous costs that are not included in the previous three categories. For example, casualty and liability insurance costs and security costs are all lumped into the administration category.

The same airport APM study (Gambla and Liu, 2012) revealed that the total annual operating cost for airport APMs in North America varies from 1 million to 15 million dollars, depending on different vehicle fleets, train consists, route lengths, and operating schedules. As documented in Table 5.4, the budgeted and actual O&M costs are fairly close, despite some applications did not include power costs in the O&M data collected. When dividing the total cost by total vehicle miles accredited, a unit O&M cost per vehicle mile may be derived for each system, ranging from $0.1 to $1.63 per vehicle mile.

TABLE 5.4   Annual O&M Cost for Airport APMs in North America

Airport APM System	Annual O&M Cost		Remarks
	FY07/08 Budgeted (M)	FY06/07 Actual (M)	
A	$ 44	—	Power not included
B	$10.93	$10.18	
C	$ 8.5	$ 8.0	
D	$ 2.23	$ 2.17	
E	$ 1.05	$ 1.02	
F	$ 1.81	$ 1.49	Power not mcluded
G	$ 2.01	$ 1.88	Power not included
H	$15.39	$13.95	Includes all O&M costs

Source: Gambla and Liu, 2012.

5.3.3 Life Cycle Cost

The Life Cycle Cost (LCC) of an automated transit system is the sum of the acquisition and support costs of all subsystems during the life span of the particular application (Anderson, 1978). The acquisition cost, often called the capital investment, generally is made of the system purchase price, plus the interest on the capital investment funding. The support cost, also called O&M cost, comprises expenses for labor, tools and equipment, spare parts and materials, and the associated logistics required to operate the system and keep operations during the useful life span.

In a given automated transit system defined by the types of components used and the services provided, the capital investment usually increases with the built-in reliability of the components and subsystems. The support cost, or O&M cost, decreases as reliability increases as the frequency of maintenance declines. As shown in Figure 5.10, the LCC exhibits the characters of a U-shaped curve with a single minimum point. Each subsystem, such as motor, controller, braking system or wayside computer also possesses a similar life cycle cost curve.

Life Cycle Cost (LCC) analysis is a method of analyzing the cost of a system or a product over its entire life cycle (Liu, Nelson, and Song 2004) with the capital and O&M costs often calculated over the useful life of the system in terms of a net present value or as a uniform annualized cost. LCC enables the researcher to define the elements included in the life cycle of a system or product, and assign equations to each element. If each subsystem is designed so that its life cycle cost is minimum, the system life cycle cost is minimum. An ideal situation is when system reliability is adequate at a corresponding minimum life cycle cost. However, the real world is often far from this ideal or perfect situation. Typically, customer service goals and

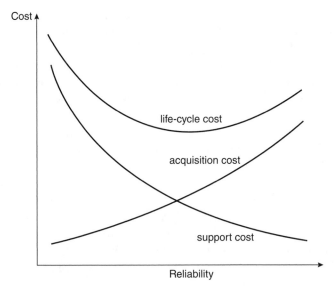

Cost

life-cycle cost

acquisition cost

support cost

Reliability

FIGURE 5.10   Life Cycle Costs for Automated Transit Applications. Data from Anderson, 1978.

safety requirements dictate that system reliability needs to be maintained or increased; therefore, the challenge becomes finding the minimum life cycle cost while maintaining the minimum acceptable level of reliability. These aspects will be explained in the next chapter.

REFERENCES

Anderson, J. E. 1978. *Transit Systems Theory. Lexington Books*, DC Health and Company, Lexington MA, Toronto.

Anderson, J. E. 1996. "The historical emergence and state of the art of PRT systems." *Infrastructure*, vol. 2, no. 1 (Fall, 1996), pp. 21–27.

Andreau, R., and J. Ricart. 2010. "Technology-driven organizational innovation at the Barcelona subway: efficiency, commitment, and firm boundaries." In 30th Annual Conference of the Strategic Management Society, Rome, 2010.

Cohen, J., et al. 2015. "Impacts of unattended train operations on productivity and efficiency in metropolitan railways." *Transportation Research Record: Journal of the Transportation Research Board*, vol 2534, pp. 75–83.

Davis, J., G. Love, and M. Sturgell. 2013. "Calculating the capacity of automated transit network systems." In: Proceedings of Automated People Movers and Transit Systems 2013: Half a Century of Automated Transit - Past, Present, and Future. Reston, VA, USA, ASCE, 2013.

Federal Transit Administration. 2010. "Annual reporting manual, 2010." Available at https://www.transit.dot.gov/ntd. Accessed in November. 2014.

Gambla, C., and R. Liu. 2012. "Airport Cooperative Research Program (ACRP) Report 37 A: guidebook for measuring performance of automated people mover systems at airports." Sponsored by the Federal Aviation Administration (FAA). Transportation Research Board (TRB) of National Academies. Washington, DC, 2012.

Graham, D., A. Crotte, and R. Anderson, 2009. "A dynamic panel analysis of urban metro demand." *Transportation Research Part E: Logistics and Transportation Review*, vol 45, no. 5, pp. 787–794.

Grava, S. 2002. *Urban Transportation systems, Choices for Communities*, McGraw-Hill, Professional Architecture.

Holle, G. 2015. "Two of a kind: Miami's Metrorail & Metromover." Miami Metro Rail. Available at http://web1.ctaa.org/webmodules/webarticles/articlefiles/Miami_Metrorail_Metromover.pdf. Accessed in March 2015.

Hsiung, S., and M. Stearns. 1980. "Phase I Morgantown People Mover impact evaluation." Report No. UMTA-MA-06-0026-80-1.

Irving, J., et al. 1978. "Fundamentals of personal rapid transit." Monograph based on a program of research 1968–1976 at Aerospace Corporation, El Segundo, CA.

Lea + Elliott. 2010. "Guidebook for planning and implementing automated people mover systems at airports. ACRP Report 37, Transportation Research Board, Washington, DC.

Litman, T. 2008. "Valuing transit service quality improvements." *Journal of Public Transportation*, 2008. vol. 11, no. 2, p. 505.

Liu, R., D. Nelson, and A. Song. 2004. "A comprehensive approach for rolling stock planning: combining train performance simulation and life cycle cost analysis." *Journal of Transportation Research Forum*, vol. 43, no. 1, pp. 105–120. Peer reviewed.

Meyer, M., and E. Miller. 2001. *Urban Transportation Planning*, 2nd ed., McGraw-Hill

Muller, P. 2015. "PRT application characteristics." Available at http://www.advancedtransit.org/advanced-transit/concept-description/. Accessed in December 2015.

Office of Technology Assessment. 1975. "Automated Guideway Transit: An Assessment of PRT and Other New Systems." Prepared at the request of the Senate Committee on Appropriations Transportation Subcommittee, June 1975. NTIS order # PB-244854.

Parkinson, T. E. 1986. "BC Transit: the world's longest fully-automated metro." *Railway Gazette International*, vol. 142, no. 2. pp. 95–97.

Paulley, N., et al. 2006. "The demand for public transport: The effects of fares, quality of 506 service, income and car ownership." *Transport Policy*, vol. 13, no. 4, pp. 295–306.

Raney, S., and S. Young. 2005. "Morgantown People Mover – updated description." In: Proceedings of Transportation Research Board Annual Conference, 2005.

Shen, L., and J. Huang. 1995. "Capacity and cost effectiveness of automated people movers." In: Proceedings of the 28th International Symposium on Automotive Technology and Automation (ISATA). Stuttgart, Germany.

UITP, 2013. "Automated metros in the world." Available at http://metroautomation .org/. Accessed in December 2013.

Watt, C., et al. 1980. "Assessment of operational automated guideway systems-airtrans (Phase II)." Monograph. DOT-TSC-UMTA-79-19 Final Report.

CHAPTER 6

ASSESSMENT OF AUTOMATED TRANSIT PERFORMANCES

It is beneficial to evaluate the success and/or failure of the existing systems before diving into massive building or heavy promotion of automated transit applications. Compared to the traditional manually operated transit systems, automated transit operation may increase transport capacity by running trains at shorter intervals and/or with other operational flexibilities (Castells, 2011). Moreover, transport capacity improvement is achieved in the safer environment of automation and with high level of economic efficiency. Building on previous studies and contemporary research, this chapter evaluates a few aspects of existing automated transit applications and presents them below.

6.1 SYSTEM PERFORMANCE

Different from highway performance, which is usually measured in vehicle miles travelled (VMT), transit performance is often measured via passenger miles travelled (PMT), which reflects the number of miles traveled by passengers reported by the system or service provider in its own fiscal year (Federal Transit Administration, 2014). Another useful way to tally transit ridership or transit use is to count the unlinked passenger trips (UPT), which is the number of boarding on transit vehicles reported by the system in its own fiscal year corresponding to the year listed. Using either a single-car

Automated Transit: Planning, Operations, and Applications, First Edition. Rongfang (Rachel) Liu.

train or long trains with multiple cars, performance of public transit, particularly automated guideway transit (AGT), may also be measured via vehicle revenue miles (VRM) or passenger car miles (PCM) that records the number of miles traveled by transit vehicles in revenue services. Train revenue miles (TRM) documents the number of miles travelled by the whole train vehicles in revenue services.

In the United States, all transit properties that are recipients of Urbanized Area Formula Grants from the Federal Transit Administration (FTA) are required to submit their operation, administration, and safety and security statistics to National Transit Database (NTD) (FTA, 2014). Established by Congress, NTD is the Nation's primary source for information and statistics on the transit systems of the United States.

There is no central authority or comprehensive data collection for automated transit operations since airport APM, the majority of automated transit applications in the United States, is not funded with federal dollars, therefore generally not subject to the performance reporting requirement. A few transit operations, such as Morgantown Group Rapid Transit (GRT) and Las Vegas Monorail did submit their performance statistics to the NTD in recent years. Among all 16 modes of transit categories included in NTD, two of them, automated guideway (AG) and monorail (MR) are directly related to automated transit applications. Since 2011, the two modes have been combined into one, monorail and automated guideway mode (MG), to represent all automated transit applications (Federal Transit Administration, 2014).

It is widely known that the transit share in the United States is rather small. Compared to the overall transit industry, automated transit operation is even minuscule. As shown in Figure 6.1, the total PMT by highway in 2010 in the United States was around 4 trillion while by transit is about 41 billion, which is about 1% of the combined PMT by surface transportation mode (Liu and Moini, 2014). The total PMT carried by automated transit in the United States during the same period is barely detectable in the transit share pie chart, about 0.03%. Given the successful implementations and rapid expansion of DLM in Asia, Europe, and other continents, it is obvious that there are great voids to be filled by automated transit applications in the United States.

Another close examination of the operation statistics for automated transit applications provides a baseline to understand the exposure, scope, and history of automated transit operations. As shown in Figure 6.2, the total PMT for AGT started around 10 million in 2002 and reached more than 21 million in the year 2012. The continuous increases in AGT ridership are composed of two main components, first is the gradual increases in AGT ridership, which peaked around 2007 for most downtown people movers (DPM). The second component of the increase is simply the result of more service providers or transit agencies submitting data to NTD. For example, Las Vegas Monorail

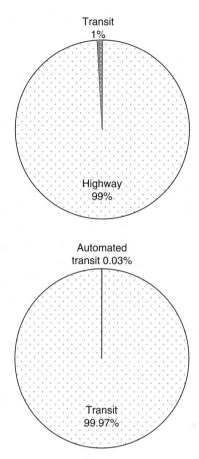

FIGURE 6.1   PMT Shares by Transit and Automated Transit USA, 2010.

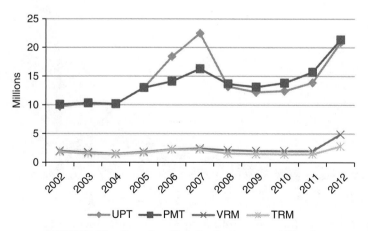

FIGURE 6.2   Operating Statistics for Automated Transit.

submitted data in the years 2006, 2007, and 2012 but not in other years included in the graph.

The total number of UPT and PMT are almost identical for AGT applications as it reflects the short trip length for most AGT riders, around 1 mile. The clear divergence between PMT and UPT for the years 2006 and 2007 reflects the short trips, less than a quarter of mile, taken by Las Vegas Monorail riders when it reported data to the NTD in those 2 years, then started again in 2012. The number of VRM and TRM are similar for all AGT systems; it is a reflection on the single- or double-cart train configuration for most of DPM and GRT systems in the United States (Liu and Moini, 2014). Only the monorail application in Las Vegas uses four-cart trains.

There is no official data collection for driverless metros (DLMs) due to its recent surge in numbers and track kilometers, as well as its scattered geographical locations. However, UITP (2013) has been keeping track of individual applications and projecting future trends. A recent report (UITP, 2013) revealed that the total track kilometers of DLM has tripled between the early 1980s and the end of 2000s. It increased six folds in the short 4-year period from 2011 to 2015, as shown in Figure 6.3.

The operating statistics for PRT is even scarcer due to the small number of applications and short existence of all three applications. Table 6.1 demonstrates the basic operating characteristics of Ultra PRT at Heathrow International Airport. Six months after its initial inauguration (Ultra Global, Inc. 2012), Ultra has accumulated a total of 4595 hours and carried over

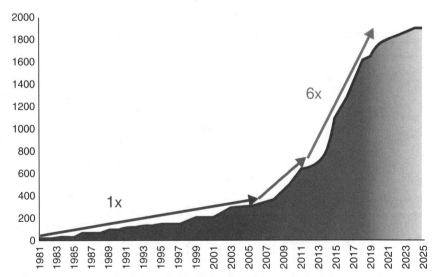

FIGURE 6.3   Track Kilometers for DLM Applications, 2013. *Source*: UITP, 2013. Reproduced with permission of Russell Publishing Limited.

TABLE 6.1   Performance Statistics for Ultra PRT in Heathrow International Airport, London, UK

Metrics/Month	September	October	November
Hours of operation	646	667	647
Number of occupied vehicle journeys	21,695	21,527	22,183
Percent of zero waiting time	86	86	84
Average waiting time (seconds)	9.6	8.3	10.3
Service availability	98.6%	99.9%	99.7%

Source: Ultra Global, 2012.

200,000 passengers with high reliability and minimal waiting times. As of early 2015, more than one million passengers have utilized the Ultra PRT in Heathrow International Airport in London (Ultra Global, Inc. 2015).

6.2 RELIABILITY

As stated in last chapter, the reliability of the transit infrastructure affects the overall expenses of automated transit applications. A few studies (Charkrabarti and Giuliano, 2015; Stewart et al., 2015; and Kuhn et al., 2014) found that reliability is highly valued by passengers as unreliability may result in unpredictable wait times, missed connections, and penalties associated with unproductive out-of-vehicle travel times (OVTT), such as waiting and walking. However, it is also true that little is known on how transit reliability affects demand, that is, it is known that unreliability discourages transit use but is not clear whether improved transit reliability is capable of lulling single occupancy vehicle (SOV) users out of their private vehicles. Furthermore, how much improvement in reliability will affect the mode choice decision results?

It is speculated (Charkrabarti and Giuliano 2015) that reliability effect may be related to the trip purposes. For example, the reliability effect appears to be stronger in the weekday peak period compared to the off-peak period. The observation is logic as most peak-period trips may be commuting or other mandatory trips, which have a predefined arrival time and location (Yang, Jin, and Liu 2008). Trips taken during off-peak period may be made of leisure or discrete purposes have less constraint in arrival times, sometime, even destinations. Therefore, it is not too farfetched to suggest that better schedule adherence can potentially promote patronage of fixed-route, fixed-schedule transit systems.

Examining reliability from the supply side, it is easily understood that reliability improvements may lead to productivity gains for transit agencies. In the case of automated transit applications, the reliability effect has been understood and taken into consideration since the early operations of Airport APM systems. Most airport operators use "availability" measures to gauge, evaluate, and compensate for their APM operators. Reliability is not directly substituted with availability but it is often measured by various availabilities, which are more tangible metrics. It is important to define a few key terms before the availability statistics are presented:

- Down time event: This measure is to count the number of down time events happened during normal operating period, which are used to measure unusual interruptions in daily operations.
- Mean time between failures (MTBF): As implied by the name, it indicates the mean time between two failures. This factor will show us how often the systems break down.
- Mean time to repair (MTTR): The time period needed to repair the systems when they break down. MTTR is a measurement to judge how well the maintenance system works.

APM system performance is usually measured as a ratio of service provided to the service required and reported as a percentage (Davis and Love, 2011). The performance measurement can be generally characterized as 'system availability' and quantifies the actual availability of the system relative to the availability specified in the O&M service contract. The main items to include in the overall APM system availability calculation are the subsystems operated and maintained by the O&M organization. Items to be included in performance measurements can vary from simply the adherence of APM trains to a specified schedule to more complex requirements, for example, including platforms, stations, platform doors, escalators, elevators, and fare gate service in the availability calculations.

A recent study (Gambla and Liu, 2012) on airport APM performance measures has categorized the airport APM service availability measurements into three tiers: A, B, and C. The simplest approach, Tier A, measures the percentage of time service has been available on the airport APM systems:

$$SA_A = AOT/SOT \times 100 \tag{6.1}$$
$$AOT = SOT - D \tag{6.2}$$

where:

SA_A: Service availability, Tier A approach
AOT: Actual operating time
SOT: Scheduled operating time
D: Downtime, the total time of all downtime events.

For airport APM applications, down time event may include the following:

- When the actual headway of in-service trains exceeds the scheduled headway by more than 20 seconds during the time when the system is scheduled to provide service.
- When any in-service train has an incomplete trip on a scheduled route during the time when the system is scheduled to provide service.
- When the first daily departure of an in-service train from the terminal on each scheduled route fails to occur within the time of one scheduled headway during the time when the system is scheduled to provide service.

If any of these downtime events occur at the same time or overlap one another, the earliest start time and the latest end time of the events are to be used in determining downtime. All downtime events are assigned to one of the following pre-defined causal categories:

- Weather-induced. Downtime caused by the weather, such as lightning striking the guideway, or a snow or ice storm, for example.
- Passenger-induced. Downtime caused by a passenger, such as a passenger holding the vehicle door open, or a passenger pulling an emergency evacuation handle on an in-service train, for example.
- System equipment-induced. Downtime caused by system equipment, such as a broken axle on an in-service train, or train control system equipment that fails while in service, etc.
- Facilities-induced. Downtime caused by the facilities, such as a station roof leaking water onto the floor immediately in front of one side of the station sliding platform doors, requiring a bypass of that side of the station, or a crack in a guideway pier that limits the number of trains in an area, etc.
- Utility-induced. Downtime caused by a utility service provider, such as the loss of an incoming electrical feed to the APM system, etc.
- O&M-induced. Downtime caused by personnel affiliated with the operations and/or maintenance organization, such as the mis-operation of the system from the control center, or the failure of a maintenance technician to properly isolate a piece of equipment from the active system operation on which he/she is working, etc.
- Other. Downtime caused by other issues, such as a terrorist threat, or delay due to the transport of a VIP, etc.

In an effort to limit the complexity of the measures and provide an alternate means of calculating service availability, some measures deliberately do not attempt to capture all system events that an airport APM user could perceive as a loss of availability. Reflecting the service reliability, Tier B approach measures the same percentage of time service has been available on the airport APM system using the following equations:

$$SA_B = MTBF/(MTBF + MTTR) \tag{6.3}$$

$$\begin{aligned} MTBF &= SOT/NF \\ MTTR &= \text{Sum of TTR/NF} \end{aligned} \tag{6.4}$$

Where:

SA_B: Service availability, Tier B approach
MTBF: Mean time between failures = Service reliability
MTTR: Mean time to restore = Service maintainability
SOT: Scheduled operating time
NF: Number of failures
F: Failure.

Similar but different from downtime, failure includes the following:

- When any in-service train has an unscheduled stoppage during the time when the system is scheduled to provide service.
- When any in-service train has an incomplete trip on a scheduled route during the time when the system is scheduled to provide service.
- When any vehicle or station platform door blocks any portion of the nominal doorway opening that passengers use to board and alight trains dwelling in station during the time when the system is scheduled to provide service.

A minor nuance is that not all delays of more than 20 seconds of the scheduled departing time were picked up or counted as failure unless it prevented the train from departure. Similarly, all failures and total restoration times are to be quantified and assigned to the causal categories, which are identical to the pre-defined causal categories for downtimes presented earlier.

Appearing the most comprehensive and most complex, Tier C approach measures the percentage of time service has been available on the airport APM by incorporating all components: mode, fleet, and station platform door into the service availability calculations:

$$SA_C = \text{Sum } SA_{TF}/\text{Sum ST} \tag{6.5}$$

$$SA_{TF} = ST \times A_{SM} \times A_F \times A_{SPD} \tag{6.6}$$

$$A_{SM} = AMOT/SMOT \tag{6.7}$$

$$AMOT = SMOT - MD \tag{6.8}$$

$$A_F = ACOT/SCOT \tag{6.9}$$

$$ACOT = SCOT - CD \tag{6.10}$$

$$A_{SPD} = APDOT/SPDOT \tag{6.11}$$

$$APCOT = SPDOT - DD \tag{6.12}$$

where:

SA_C: Service availability, Tier C approach
SA_{TF}: Time-factored service availability value
ST: Service time of each service period, in hours
A_{SM}: Service mode availability
AMOT: Actual mode operating time
SMOT: Scheduled mode operating time
MD = Mode downtime
A_F = Fleet availability
ACOT: Actual car operating time
SCOT: Scheduled car operating time
CD: Car down time
A_{SPD} = Station platform door availability
APDOT: Actual platform door operating time
SPDOT: Scheduled station platform door operating time
DD: Door downtime.

At the first glance, one can be easily intimadated by those many equations and variables. But once examined closely, it is not difficult to discover that all three tiers measure the percentage of service time provided, as specified earlier. The data collection need is also identical and the central control for automated operations made it easier to collect the data on a daily, weekly, or monthly basis before further analyses or data manipulations are carried out.

Using terms defined above, many automated transit operation agencies, especially airport APM operators, specified various availability measures, such as mode availability, station availability, platform availability, fleet availability, service availability and/or system availability, to gauge the service quality of automated transit operations. System availability has a very broad definition and varies among different applications (Davis and Love, 2012). For example, both the SkyLink APM in Dallas-Fort Worth International

Airport (DFW) and the AirTrain APM application in San Francisco (SFO) International Airport included mode, fleet, and station availability in their service availability measures, the Tier C approach. The Satellite Transit System in Seattle–Tacoma International Airport (SEA) uses the combination of MTBF and MTTR to gauge the service available, Tier B approach. The APM system in in Tampa International Airport (TPA) utilizes the simple ratio of actual operating time and scheduled operating time to measure the system availability, a Tier A approach.

The availability measures listed above are in wide usage in airport APMs worldwide. According to current performance measures, the APM operator or contractor will receive full compensation when the APM trains are available to a predefined level of availability. On the positive side, the connections between system availability and compensation rate for APM operators serve as incentive for smooth and reliable APM service, which is a necessity for airport operations. The negative impact of such connection may results in lack of motivation to optimize APM operations, especially in airports.

For example, many operation plans for APM applications were developed before their inauguration, and few have been adjusted many years after their operation. There is no incentive for APM operators to match their operation supplies to passenger demand, which may have experienced dramatic changes in the past decades in addition to normal daily, weekly, and seasonal patterns. The system availability measure provides an appropriate level of accountability for the system and its elements that most directly affect the quality and level of service experienced by passengers. What is lacking in the current performance measures is efficiency, balance between supply and demand, and customer satisfaction.

Most automated transit applications are designed for very high service availability, especially those newly established, small-scale operations. Figure 6.4 demonstrates how monthly availability was recorded and calculated by an airport APM operator, who derives penalties and/or bonuses for the APM contractor based on the availability metrics each month.

Given the nature of automated operation and short operating history of most automated transit applications, most automated transit systems maintain very high availability and reliability records. The availability rate may drop while system or the components of the system age. For example, during the first 5 years of Morgantown GRT operation, the system availability for the five-station system steadily rose from 95% and hovered near 99% until the late 1998 when a declining trend began. The average system availability declined to around 97.5% in 2006 and it has remained below 98% since then, despite improvements to the computer system (Gannet Fleming, 2010).

Many components or sub-systems of the Morgantown GRT need to be replaced or improved after nearly four decades of operation. For example, the

Monthly performance report

Month	8/2007
Monthly report no.	DCC-OpsM.466
System1	1422 h 38449 km
System2	1402 h 37220 km

Month	TTR	No. of failures	Total no. of stoppages	MTTR	MTBF	Availability
6/2007	4,95 h	7	33	0,71 h	170,72 h	99,59 %
7/2007	5,79 h	14	40	0,41 h	88,16 h	99,53 %
8/2007	11,51 h	15	50	0,77 h	81,9 h	99,07 %
Average	22,25 h	36	123	0,62 h	101,6 h	99,4 %

Overall performance	Total operating hours	Total operating hours				

Stop breakdown		Stops	Time
	Technical	98	6,11 h
	Operational	20	2,58 h
	External	5	0,65 h
	Partial	24	26 h

Operability	
Total operating hours	3679,97 h
Total downtime (all stoppages)	31,58 h
Operability %	99,14 %

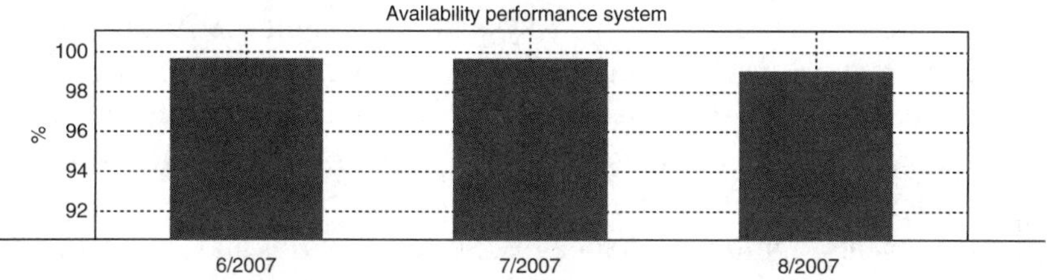

FIGURE 6.4   Sample System Availability Report.

original minicomputers were discontinued, resulting in spare parts shortages. The DEC computer hardware was replaced by Dell PCs in 1999 with an accompanying port of the assembly language software to the C language. Before porting to the C programming language, the control software consisted of 75,000 lines of PDP-11 assembly language code. Guideway snow-melting heating piping had to be replaced. Power rail had to be replaced. With the Morgantown GRT's declining system availability, West Virginia University's leadership recognized the need to assess, maintain, and improve the primary and critical subsystems in a systematic manor (Gannet Fleming, 2010). The problems Morgantown faced with obsolescence provide an important lesson for today's APM or GRT developers.

In a recent study, Cohen et al. (2015) measured the reliability of DLM using mean distance between failures (MDBF). In this case, the distance is defined by million train kilometers (MTKM) and incidents greater than 5-minute delays. As noted in the last section, passenger-focused metrics such as train or passenger delay hours are preferred but often not available. In this case, the MDBF data have been checked, verified, and therefore presented here. As shown in Figure 6.5, the DLM applications all have reliability levels within the top one-third of their respective network wide averages.

6.3 SAFETY AND SECURITY

One of the promises of automated transportation system in general is the safety improvement achieved by adaptive cruise controls, collision avoidance, eliminating human errors all together when the human drivers are augmented or replaced by precision mapping, sensing, detecting, and guiding systems. For automated transit applications, the elimination of human drivers has the potential to avoid human error and working fatigue all together when the routine navigation and operation of transit trains are delegated to computers and communication networks.

As early as in the 1970s, systems safety and passenger security (SS&PS) procedures have been developed for the assurance of actual and perceived passenger safety and security in AGT systems (Benjamin, 1979). In conventional transportation systems, transportation personnel can help to evacuate and rescue passengers. AGT systems, however, because of their unmanned nature and unique configurations, present a number of challenges related to evacuation and rescue. Operation of AGT applications with elevated guideways also presents significant challenges. Serious injuries and loss of life can result from situations in which inadequate means of evacuating and rescuing passengers exist. There were theoretical safety and security procedures, such as evacuation and rescue procedures built in place for all the existing automated transit applications, but the small operations and scarce occurrences of emergency events made it almost impossible to test those procedures.

A dilemma faced by the transportation professionals and automation advocates is that there is no application in the areas of automated vehicles and very limited safety records for automated transit systems. Therefore, the limited safety and security data collected by existing AGT systems since the new millennium become the only base for safety evaluation and comparison of automated transit applications.

6.3.1 Safety Records for Automated Guideway Transit

As one of the newest segment in the NTD, safety and security module was created for gathering data from various transit agencies starting 2002 (FTA,

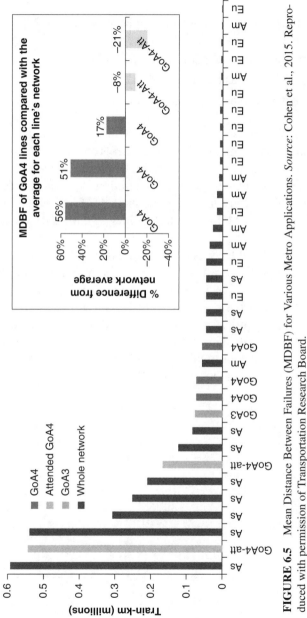

FIGURE 6.5 Mean Distance Between Failures (MDBF) for Various Metro Applications. *Source:* Cohen et al., 2015. Reproduced with permission of Transportation Research Board.

2014). Using S&S-40 Form, transit agencies generally collect and submit incident, fatality, and injury information to NTD on a monthly basis. There are four different subcategories under the general incident category:

- Collisions: includes all collision types reported to NTD, excludes suicides
- Derailments: includes all derailments reported to NTD
- Fires: includes all fires reported to NTD
- NOC: includes all other reportable incidents.

For evaluation purposes, a study (Liu and Moini, 2014) included collision, derailment, and fire as incidents. These three forms of incidents are considered as the system's malfunction and operation deficiency, since other forms of incidents are not pertinent to the systems' operation. Also, after NTD changed the classification of AGT from 2012 to integrate AGT and Monorail (MO) in one classification called MG (Monorail/Guideway), the safety and security metrics are combined also.

Figure 6.6 shows that there was no passenger fatality by AGT since its inauguration in the 1970s. There was usually no major incident reported by AGT applications in the United States from 2002 to 2012 except in the years 2005, 2009, and 2012, when two, two, and one incidents were reported, respectively. Similarly, passenger injuries rarely occurred within AGT operations except in the year 2005 and 2010. The majority of the 18 passenger injuries in 2010 were caused in one incident, which took place in Metromover, Miami, FL (National Transportation Safety Board, 2012).

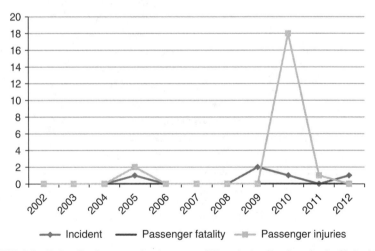

FIGURE 6.6   Safety Performance by Automated Transit Applications in the United States, 2002 –2012.

According to the National Transportation Safety Board (2012), on July 20, 2010, an inbound Miami-Dade Transit (MDT) Metromover train, traveling about 10 mph along a fixed guideway, struck the trailing end of another Metromover vehicle. The struck Metromover was stopped at Brickell Station near downtown Miami, Florida. There were a total of 45 passengers on board the two Metromover trains. These Metromover vehicles operate in a fully automatic mode without human operators. Sixteen passengers incurred minor injuries and were transported to, treated by, and released from local hospitals. Total damages were estimated at $406,691.

The National Transportation Safety Board determined that the probable cause of the accident was the Miami-Dade Transit rail traffic controllers' decision to restart automated train operations without accounting for the location of all Metromover vehicles following a safety shutdown after the signal rail had been damaged by a defective Metromover guide wheel. Contributing to the accident was inadequate oversight by Miami-Dade Transit. The event demonstrated that human error is not limited to train cabs or transit guideways, but it may also occur in control centers.

It is easy to contribute the zero fatality and small number of injuries and incidents to the small operation scale of automated transit applications. Studies and practices indicate that crash frequency or the total number of fatalities and injuries, are not apt measurement metrics, as they do not consider crash or risk exposure. Similarly, the low total fatality and injury numbers in overall transit systems are often brushed off by small shares of transit use in the United States.

6.3.2 Comparison with Other Guideway Transit

In order to account the risk exposure and bring the safety and security records into proper context, the accident rates should be normalized via operation scales.

Crash exposure is defined according to the objective of analyses and modes of transportation under study. For instance, in highway safety studies, high crash locations are commonly identified by normalizing crash frequency against traffic volume, such as annual average daily traffic (AADT).

To avoid the biases created by the risk exposures or uneven shares of usage, safety performance measures for multiple modes of passenger transportation are usually normalized by utilizing various platform units, such as per million passenger miles travelled (MPMT), per million unlinked passenger trips (MUPT), per million vehicle revenue miles (MVRM), or per million train revenue miles (MTRM).

From operational perspective, the exposure is captured through VRM and train revenue miles (TRM), which are used here to normalize fatality, injury,

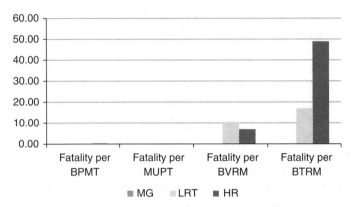

FIGURE 6.7 Passenger Fatality Rates.

and incident frequencies. Both VRM, the number of miles traveled by transit vehicles in revenue service, and TRM, the number of miles traveled by whole train vehicles in revenue service, reflect transit system's operation and availability of resources. Thus, the normalization of incidents, injuries, and fatality against these factors reflects the level of operational safety. It is helpful to note that the normalization is performed using "million" PMT, UPT, VRM, and TRM for injury and incident rates and "billion" for fatality rates to achieve better readability and presentation.

As depicted in Figure 6.7, when measured in fatalities per billion passenger miles traveled (BPMT) and billion UPT (BUPT), AGT with zero passenger fatality is the safest mode of transportation among fixed guideway transit systems during the period of 2002–2012. The average fatality rate is 0.4 per billion passenger mile travelled (BPMT) for LRT and 0.3 per BPMT for subways during the same 10-year span. Overall fatality rates per BPMT for all three modes are very low as indicated by the barely visible two clusters in the left side of Figure 6.7.

The fatality rates per billion vehicle revenue miles (BVRM) are significantly higher for both LRT and heavy rail even though it remains zero for automated transit during 2002–2012. The magnitudes of fatality rates per BVRM and BTRM for all three modes are depicted by the two clusters in the right-hand side of Figure 6.7. The fatality rates per billion train revenue miles (BTRM) are further magnified for both LRT and subway as they usually employ longer trains than automated transit. The increase is almost five folds for subway, which is consistent with its longer trains with 6 to 12 cars.

The injury rate for automated transit, when normalized via million PMT (MPMT), is actually highest among all three modes, which is largely the result of the single accident occurred in Metromover, Miami, in 2010. As shown in Figure 6.8, injury rates for subway is actually lowest when total

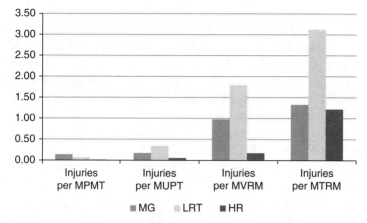

FIGURE 6.8   Passenger Injury Rates.

injuries are distributed among MPMT, MUPT, MVRM, and MTRM. This shift may be explained by the isolated environment for subway applications, which is important to reduce conflicting points and therefore injury numbers, especially when compared with LRT, which often operates in mixed traffic. The second reason for subway's lowered injury rates is due to its relatively larger operation scale, which derives a lower rate when the number of injuries is distributed among larger PMT, UPT, VRM, and TRM units.

When 10-year average incident rates are distributed via MPMT and MUPT, subway faired the best with an average 0.01 incidents per MPMT and 0.03 incidents per MUPT, as demonstrated in Figure 6.9. Subway retained its first place when incident rates are measured against MVRM and it only lost its

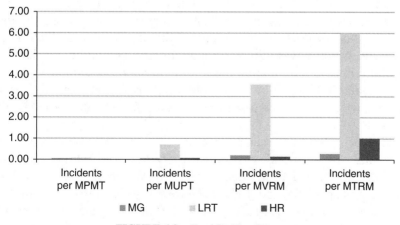

FIGURE 6.9   Total Incident Rates.

place to automated transit when incident rates are measured via number of incidents per MTRM due to the long trains subway usually operates. With all four measures, incident rates per MPMT, MUPT, MVRM, and MTRM, LRT retained the highest incident rates for all four categories. The most likely cause for LRT's high incident rates is probably the mixed operations with other traffic. Running on street levels or sharing track with other commuter or freight rail may increase conflicting points, which made LRT operation more susceptible to interruptions and incidents. The relatively small-scale operations for LRT do not help either.

Several general trends may be observed from the incident, injury, and fatality rates by each mode. First, the total number of incidents, injuries, and fatalities for all three modes fluctuate among different years, except the passenger fatality for AGT remains zero. When converted into unit rates per BPMT, BUPT, and BVRM, it is clear that LRT has the highest accident, injury, and fatality rates than subway throughout the past decade while AGT has the zero passenger fatality, smallest injuries, and least amount of incidents. It is only when fatality are measured via million TRM, subway turns out to be the highest due to its long trains with multiple cars.

Second, the total number of accidents per year is generally on the rise largely due to the gradual increase of ridership for all three modes. The acceleration rates for AGT is the smallest as there was no system expansion during the past decade and the total increases were mostly caused by more properties started to report safety and security data. The acceleration rate for LRT is the largest as the increase in incidents and injuries outpaced its increase in PMT and other operation metrics.

Transit incidents, fatalities, and injuries vary considerably from year to year, especially when viewed alone, without consideration of the volume of ridership. However, the true picture of transit safety for various modes starts to emerge when the simple accident or fatality numbers are examined in the context of operations. Being presented as annual total amount, the total fatality for subway has been consistently higher than that of LRT during the past decade, which may gave the wrong impression that subway was less safe than LRT. However, when examining the total service magnitude, such as PMT, the total PMT for subway was often four to six times of LRT for the same year, which helps to explain the much lower fatality rates by PMT, UPT and VRM.

Putting the safety performance measures into the proper context, that is, normalizing with operation statistics, such as PMT or VRM, it is clear that the accident rates, such as accident rate per MPMT, injury rate per MPMT, and fatality rate per BPMT for LRT are consistently higher than those of AGT and subway. The normalization is achieved by converting the total number of accidents, injuries, and fatalities into two large categories, passenger exposure and operation exposure, which are measured by PMT and UPT for the former as well as VRM and TRM for the latter.

The general observation of guideway transit safety is also consistent with general expectations of intermodal safety performance comparison; that is, AGT and subway usually operate in exclusive right-of-way, which have less exposure to passenger and operation conflicts while LRT usually operates in mixed traffic, which has more exposure to various conflicts and susceptible to accidents. Furthermore, with lighter and smaller vehicles, narrower guideway foot print, and fully automated central control systems, AGT faired the best in all aspects of safety performances.

6.4 COST-EEFECTIVE ANALYSIS

As presented in the previous chapter, the total capital investment, operations and maintenance expenses, and even the life cycle cost of a particular transportation project may not tell the whole story. The high investment costs are often associated with large-scale and long-lasting projects, which may bring great benefit, and less expensive projects may provide temporary solution or mandate continuous incremental investment. Of course, there are exceptions to those general rules; therefore, cost-effectiveness analysis (CEA) is often employed to estimate and compare various alternatives.

CEA is a technique that relates the costs of a program to its key outcomes or benefit (Cellinin and Kee, 2010). Cost–benefit analysis may have the potential to bring CEA one step further to compare costs with the dollar value of all or most of a program's many benefits but the process of conducting a CBA or CEA is more complicated than it may sounds from a summary description or a direct accounting practice. In the case of automated transit applications, the capital and O&M costs maybe concrete while some of the benefits may be tangible, direct or immediate but others intangible, indirect and sometime, elusive or take a long time to manifest.

For the capital investment on DLM conversion or new installations, a recent study (Cohen et al., 2015) documented the additional costs of DLM equipment reported by seven metros located in North and South America, Asia, and Europe. The study found that "additional cost" for DLM depends on the baseline to which equipment is being compared, that is the technologies and capabilities used by a metro's existing conventional systems. For example, Asian metros tend to install platform doors as standard, therefore; platform doors are not considered "additional" for Asia DLM applications.

The basic levels of standard rolling stock also affect the "additional costs" of converted DLM operations. As presented in Chapter 5, the metro, which reported 20% "additional cost" in rolling stock, had basic bare bone trains to start with. For others, such as those reported "same" or minimum increase in rolling stock costs, the standard train for conventional lines already have advanced passenger information systems and remote condition monitoring capabilities that are the key element of DLM operations.

Consistent or comprehensive data on operation and maintenance cost of automated transit systems is not available largely because the newly developed metros are often part of conventional metro network and transit agencies did not disaggregate the costs between automated and conventional lines (Cohen et al., 2015). For airport APMs, which are often aggregated within the airport operations, the O&M cost data is not easily accessible since no organization is required to gather or submit such data to the government (Gambla and Liu, 2012).

Selected surveys revealed that DLM operations may reduce staff numbers by 30–70%, while the amount of wage cost reduction depends on whether staff on DLM lines is paid more. Anecdotal evidence from two metros indicated that it is not possible to translate staffing levels directly to operational costs, because multi-skilled staff on DLM lines may be paid more than drivers or station staff on conventional lines (Cohen et al., 2015). Wage cost changes would also be affected by a metro-specific decision about headcount reduction and any average wage increase.

The same study (Cohen et al., 2015) reported that maintenance costs are also difficult to quantify. For example, one metro in Europe and two in Asia indicated that remote condition monitoring and equipment redundancy reduce the amount of preventive maintenance on DLM lines, but also noted that spare parts are more expensive. There are two metros that documented maintenance cost differences between automated and conventional lines but it is impossible to identify the causes due to the differences in organizational or technological factors.

After presenting the basic components of life cycle cost (LCC) of transit system in the last chapter, it is natural to point out that capital cost, O&M cost, or LCC alone will not complete the cost evaluation criterion. The fourth dimension, user allocation, is needed to measure the cost-effectiveness of the guideway transit, especially AGT systems. As explained earlier, a large number of APM applications are evaluated against its availability alone, which may be part of the reason that many empty APM trains circulate the airport, downtown or major activities centers during off-peak hours. To promote efficiency and effectiveness, it is important to match supply and demand— train operates because there are riders using the services, not simply because contractors need to meet their service availability standards.

Allocating the initial capital cost to riders who use the automated transit services throughout the life span of the facility, the concept of user allocation of annualized capital cost (UAACC) is introduced here. That is, the initial

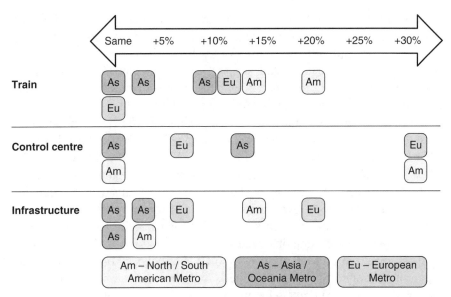

FIGURE 6.10   Additional Capital Costs of Technology for DLM Applications. *Source*: Cohen et al., 2015. Reproduced with permission of Transportation Research Board.

capital costs are converted to annualized costs based certain number of years of the project life span, say 40 or 80 years, assuming a certain interest rate, such as 6% or 8% per year. The conversion factors used may be specific to a particular project since the life span of individual elements of the rail system vary and interest rate changes from period to period: however, the concept of the user allocation is the key, which provides a levelled playfield for all modes via a basic platform, per user or per passenger mile travelled (PMT).

As an example, Figure 6.10 illustrates how the user allocation of annualized capital costs (UAACC) is compared among AGT, LRT and subway also called Heavy Rail (HR) modes. The PMT for both heavy and light rails were collected in a certain year and used as an indicator for rail usage. In the case of AGT or particularly for APM, since not all PMT were collected, especially those systems located in airports, an equivalent place miles traveled (PLMT) was substituted and presented. The PLMT is derived by multiplying the capacity of the vehicle with the vehicle miles traveled. To bring the APM to a realistic comparison with other two modes, the PLMT was computed using half of the system capacity. The assumption is that the vehicle capacity is used about 50%, which is close to the PMT/PLMT ratio of Metromover in Miami and a few other places.

The UAACC possesses several advantages in cost comparison. First, the magnitude of the cost is much closer to daily transit operations in terms of dollars and cents, it is easier for decision makers, planners, and the general public to relate and compare. Second, the smaller magnitude of the user allocation is more sensitive to system changes than the total cost; therefore, the differences in individual systems are clearly revealed. Third, by linking the travel demand, PMT, to the system supply, capital cost; the UAACC measures the cost-effectiveness of the system, which should be the ultimate judgement on rail transit, automated or manually operated applications.

A brief observation that may also help argue the case for APM applications is that the UAACC for AGT modes are categorically lower than the other two, LRT and HR, modes. At the rate of $0.34 per PLMT, the UAACC of APM is less than half of that for LRT, $0.94, and merely a quarter of that for Heavy Rail, $1.52. Another way to state the meaning of this metric is that the UAACC for AGT is still lower than LRT even if only a quarter of the AGT vehicle seats are occupied by passenger, which is hardly the case, especially in busy airports as we often observed.

The comparison presented in Figure 6.10 is only based on a few applications around United States with limited, sometimes dated, data. For further research, there are two immediate undertakings. First, it should be very helpful if more data can be collected and meaningful statistical analyses can be performed. Second, the parameters used in the evaluation, dominated by the data availability, should be expanded and validated to refine the UAACC and evaluation process.

It is noted that technology, location, inflation/time value all had great impact on the capital cost of individual transit systems. However, the picture is not complete until the fourth dimension, user allocation, is added. It is hoped that the criterion presented here will be a starting point for later data collection and accumulation of guideway transit operation and investment data collection. In order to improve the comparability of various applications, a database containing accurate and abundant information should be established and maintained so future cost estimates will be more "scientific" than "guesstimates." The projected cost will be close to the real cost of each project.

REFERENCES

Benjamin, D. 1979. "Evacuation and rescue in automated guideway transit, vol 1: data collection, scenarios, and evaluation." Monograph. DOT-TSC-UMTA-79-47-I Final Report.

Castells, R. 2011. "Automated metro operations: greater capacity and safer, more efficient transport." *Public Transportation International*, vol. 60, no. 6, pp. 15–21.

Cellinin, S. R., and K. E. Kee. 2010. "Cost effectiveness and cost-benefit analysis." In: *Handbook of Practical Program Evaluation*, 3rd ed., edited by J. S. Wholy, H. P. Hatry, and K. E. Newcomer. John Wiley & Sons.

Chakrabarti, S., and G. Giuliano. 2015. "Does service reliability determine transit patronage? Insights from the Los Angeles Metro bus system." *Transport Policy*, vol. 42, no. 1, pp. 12–20.

Cohen, J., et al. 2015. "Impacts of unattended train operations on productivity and efficiency in metropolitan railways." *Transportation Research Record: Journal of the Transportation Research Board*, vol. 2534. pp. 75–83.

Davis, J. H., and G. W. Love. 2012. "Measuring the performance of automated people mover systems." In: Proceedings of Automated People Movers and Transit System 2011. America Society of Civil Engineers.

Federal Transit Administration. 2014. "Annual Reporting Manual, 2014." Available at https://www.transit.dot.gov/ntd. Accessed in November 2014.

Gambla, C., and R. Liu. 2012. "Airport Cooperative Research Program (ACRP) Report 37 A: guidebook for measuring performance of automated people mover systems at airports." Sponsored by the Federal Aviation Administration. Transportation Research Board of National Academies. Washington, DC.

Gannet Fleming in Association with Lea + Elliott et al. 2010. "PRT Facilities Master Plan." Prepared for West Virginia University. June 2010.

Graham, D., A. Crotte, and R. Anderson. 2009. "A dynamic panel analysis of urban metro demand." *Transportation Research Part E: Logistics and Transportation Review*, vol. 45, no. 5, pp. 787–794.

Kuhn, B., et al. 2014. "Effectiveness of different approaches to disseminating traveler information on travel time reliability." The Second Strategic Highway Research Program (SHRP 2) Report S2-L14-RW-1. Transportation Research Board (TRB), National Academy of Science. Washington, DC.

Liu, R., and N. Moini. 2014. "Benchmarking transportation safety performance via shift-share approaches." *Journal of Transportation Safety and Security*, vol. 7, no. 2, pp. 124–137.

Liu, R., and N. Moini. "Proven safety advantages of automated guideway transit systems." In: Proceedings of Transportation Research Board (TRB) 94th Annual Meeting, January 2015. Washington, DC. Peer reviewed.

National Transportation Safety Board. 2012. "Railroad accident briefing." Available at http://www.ntsb.gov/investigations/AccidentReports/Reports/RAR1101.pdf. Accessed in June 2015.

Richards, L. G., I. D. Jacobson, and R. D. Pepler. 1980. "Ride-quality models for diverse transportation systems." *Transportation Research Record: Journal of the Transportation Research Board*, vol. 774. ISSN: 0361-1981.

Stewart, A., J. Attanucci, and N. Wilson. 2015. "Comparing ridership response to incremental BRT upgrades in North America considering demographic and network effects." In: Proceedings of 2015 Transportation Research Board Annual Conference, Washington, DC, January 2015.

UITP. 2013. "Observatory of automated metros world atlas report, 2013." Available at http://www.uitp.org/sites/default/files/cck-focus-papers-files/Annual-World-Report-2013.pdf. Accessed in June 2015.

Ultra Global. 2012. "Heathrow performance statistics." Available at http://www. ultraglobalprt.com/wheres-it-used/heathrow-t5/. Accessed in December 2015.

Ultra Global Inc. 2015. "Ultra provides environmentally sustainable 21st century transport solutions." Available at http://www.ultraglobalprt.com/about-us/. Accessed in March 2015.

Watt, C. W., et al. 1980. "Assessment of operational automated guideway systems-airtrans. Phase II." Monograph. DOT. DOT-TSC-UMTA-79-19 Final Report.

Yang, F., X. Jin, and R. Liu. 2008. "Tour-based time-of-day choices for weekend non-work activities." *Transportation Research Record: Journal of the Transportation Research Board*, vol. 2054, pp. 37–45. Peer reviewed.

CHAPTER 7

PLANNING CONSIDERATIONS

A few feasibility studies conducted in recent years (Carnegie and Hoffman, 2007; Page, 2012; and Furman et al., 2014) concluded that the technology for automated guideway transit (AGT), especially personal rapid transit (PRT) or automated transit networks (ATN) is mature but market, not ready. If our European and Asian counterparts have been operating the automated transit applications for many years with efficiency and without any incident, why America could not replicate the experience? One of the suggestions is that our planners do not know how to forecast the travel demand for the AGT services based on traditional travel demand models.

Searching for solutions rather than pointing fingers, it is helpful to understand how the transit planning process relates to the culture of the transit industry. As many planners and transit executives know, planning is all about change, going into the unknown, and taking risks. However, transit riders expect the exact same service every day. Transit riders are upset by change and the resulting complaints do not reflect well on transit executives or staff. It is a challenge to convince transit agency boards and executives to undertake new services and an even greater challenge to get them to accept a new and relatively unproven mode.

According to Furman et al. (2014), ATN does not appear "on the radar" of urban planners, transit professionals, or policy makers when it comes to designing solutions for current travel problems in urban areas. Most urban

Automated Transit: Planning, Operations, and Applications, First Edition. Rongfang (Rachel) Liu.
Copyright © 2017 by The Institute of Electrical and Electronic Engineers, Inc. Published 2017 by John Wiley & Sons, Inc.

transportation project planners, developers, and policy makers are generally not aware of ATN and its potential benefits, tradeoffs, and implications. Nor are they aware of the current state of potential suppliers, whether there is a market for ATN, and what is entailed in planning, procuring, and funding ATN or automated transit systems (ATS). As the United States contemplates the future of highway infrastructure, measures out a sustainable energy future, and accommodates historic demographic shifts back to growth in urban cores, it is vital to develop a general procedure to include automated transit applications in multimodal transportation investment plan.

Even when transit agencies were made aware of the new technology or service alternatives, there were several key reasons that PRT or automated transit application in general was not included in the alternative analysis (AA) processes. First, there is a need for extensive system design, development, and test for any new automated transit applications in order to prove their safety and reliability. In reality, many transit executives have been approached by proposers that do not consider such test a necessity. It is vital to build full-size vehicles and a full-scale testing track, which includes all the important design features and run it continuously through all four seasons to know how it will perform. To operate the testing facility, a sufficient number of highly skilled, highly compensated testing engineers are needed on the payroll. Risk management and insurance issues should also be considered. With many unknowns, it is very difficult, if not impossible, to develop an accurate budget and/or timeline for testing and improving processes.

Second, there was no reliable cost data for construction or operations. For intermodal transportation systems, especially, transit services, it is important to estimate or forecast the demand, the need for accurate cost estimates is even more important from a planner's perspective. The limited data available for automated transit applications in the United States have pointed to the fact that all actual construction costs have far exceeded their respective estimates. Of course, it was true that almost all rail modes including light rail transit (LRT) and subway applications have exceeded the estimated costs but automated transit technology had far fewer examples from which to draw experiences or base comparisons.

The third issue is that many promoters of PRT were peddling proprietary systems. The transit industry has found that buying into a proprietary system, for which only one supplier exists, can lock the agency into a situation where that supplier can charge anything it wants for replacement parts, new vehicles, and guideway extensions. For example, The Port Authority of New York and New Jersey (PANYNJ) had to go through difficult negotiations with the monorail vendor to agree on a price when Newark Airport Rail Station was designed and developed.

The last but not least planning issue associated with the lack of automated transit application in the United States is the impact of National Environmental Policy Act (NEPA). For each major investment project, the transit agency or owner of the project is required to prepare an Environmental Impact Statement (EIS) that considers many factors and involves developing the concepts under open public and political scrutiny. Elevated guideways and stations are unpopular. Citizens often exert political pressure on elected officials, which in turn influences the transit planning processes. As a result, the automated transit application was often eliminated in the early stages of planning even if it was squeezed in during the initial evaluation stages.

As an accumulative result of various reasons mentioned earlier, there is no automated transit application in urban areas that people could look at and ride in the United States. It is observed that public officials had no understanding of transit modes like LRT, PRT, and bus rapid transit (BRT). They need to see real systems before they would buy into the planning process. It is possible to bring people to places like San Diego, Los Angeles, and Portland to ride LRT systems, and Pittsburgh and Ottawa to ride BRT, and to talk with local officials. There is not a model AGT application for public officials or general public to ride and experience. It is true that a number of automated people mover (APM) applications are operating at many airport terminals, that context is very different and not comparable to urban area transit operations. The only close approximation of PRT applications in the United States is the Morgantown group rapid transit (GRT), which has been implemented in urban area, even dominated by a university campus, but it is also fraught with cost overruns, schedule delays, and technological development problems, to the point that it would not serve as a compelling example.

Many researchers and scholars developed artist's renderings and computer simulations to demonstrate what different transit modes would look like and how they would perform (El-Aasar, Willard, and Hibbs, 2006 and Zheng and Peeta, 2014). Without the backing of a real operation, those simulations or artistic renderings do not lend much credibility. There was wide-spread skepticism on the part of decision-makers and the public regarding visual techniques. Bombarded with too many artistic renderings and computer simulations from those who do not have any authority or credibility in the areas of transit operations, many public officials or decision-makers are repelled from any visual presentations or evaluations of futuristic development.

There are many unanswered questions about automated transit, especially PRT applications, it also raises the question whether PRT is a mature technology or still in the womb of research and development. There is less doubt about the technology maturity of APM and driverless metros (DLMs) due to the wide-spread and large-scale applications around the world. The following sections will try to address the planning issues from three distinct but related perspectives: public policy, long-range planning, and operations planning.

7.1 PUBLIC POLICY

As described in the early chapters, automated transit technologies have been inaugurated in the large historical background of "landing on the moon" and then newly established Urban Mass Transportation Administration (UMTA) in the federal government in the 1960s and 1970s. By the same token, the national needs of shifting military capacity to civil use have resulted the over design, over built, and over promises of AGT or GRT applications, which have contributed to cost overruns and eventually the negative perceptions of AGT applications in the United States. Learning from the past experiences, the important questions that need to be answered are:

- Who are the stakeholders of automated transit development and what roles those individual entities should play?
- What policies should be developed to facilitate the development of automated transit technology and application in order to improve the overall efficiency of transportation systems?

After the demonstration projects in the 1970s and 1980s in the United States, there was no institution at the national and/or international level that is directly involved in the automated transit technology development. The few organizations that have continued their interests in the automated transit development, such as American Society of Civil Engineers (ASCE), Transportation Research Board (TRB), and European Public Transportation Commission (UITP) are often limited to standard specifications, information dissemination, and data collection.

7.1.1 Research

The ATS Committee designated as AP040 by TRB has been leading the way to define and shape the research agenda for ATS around the world (Liu, 2013). During its early years, from 1998 to 2012, the committee focused on major activity centers (MAC) circulation systems, airport automated people movers (AAPM), and PRT. The larger transportation community often turns to this committee for research development and project expertise related to the AGT technologies.

With the rapid development in ATS, especially DLM in Europe and Asia, the committee's historical experiences, technical expertise, and professional networks assembled and represented by its members become more predominant and critical. Since the turn of the new millennium, the scope of the AP040 committee is no longer limited to a few selected institutions or confined proximity of MACs. The committee members have lent their expertise to many communities in the United States, such as San Jose, Ithaca, Denver, and Phoenix, which explored the potential of ATS in order to integrate their transportation and land use development.

The Observatory of Automated Metros is another international organization commissioned by UITP, the association of public transportation agencies in Europe, to disseminate and share current and relevant knowledge about DLM (UITP, 2013). As the counter part of APTA in Europe, UITP publishes an Annual World Report on automated metros based on information obtained from industry press and/or directly from the operators and authorities who manage the automated lines. The organization also collects data on various plans and attempts to forecast future trends in the areas of automated transit development.

There are many other individuals and small businesses, which are enthusiastic about the technologies of AGT, service qualities of PRT, or purely the futuristic images of automated transit applications. Those individuals or institutions volunteer as the custodians of wealthy information on the AGT history, technology, and applications. However, since there is no systematic effort or sustainable financial backing, the information and excitement tends to be lost when an individual retire, decease, or leave the field. A few garage entrepreneurs who have developed protocol vehicles or operated small-scale testing tracks were often passed for government funding as most national grant favors large institutions with long history of operation and full line of technical expertise.

7.1.2 Design Standards

Realizing the benefits of standardization to organizations that specify and procure APM systems, such as regulatory authorities, system suppliers, system operators, system users, and the general public, ASCE has taken the lead in developing "Automated People Mover Standards" (Committee of Automated People Mover Standards, 2006). The ASCE standards include minimum requirements for design, construction, operation, and maintenance of APM systems, especially on the subject of the physical operating environment, system dependability, automatic train control, and audio and visual communications. The ASCE standards have no legal authority in their own right and have not been adopted by any authority that has jurisdiction over

AGT applications; nevertheless, they serve as a general guideline for transportation professionals to plan, build, and manage AGT systems in the years to come.

There is clearly a void in terms of definitions of various automated transit applications, their anticipated functions, and associated design specification in the past. For example, a group of citizens in Atlanta, GA was asked to evaluate a PRT application by seeing a picture of 12-person AGT vehicles (Anderson, 2009). The citizen's response was that the vehicle was too big as they did not want to ride with others. The over design of guideways and vehicles in the early stage of AGT development may have stemmed from the safety concerns of transportation engineers but also caused by the lack of understanding of automated transit applications and their respective functions. It is possible to avoid similar mistakes if individual sub-modes of automated transit applications presented in this book will be clearly defined and the design standard, such as that by ASCE, are closely adhered.

7.1.3 National Policy

Given the complexity of automated transit technology and intensity of capital investment, it is critical to have support at national /international levels for automated transit applications to take place. The support may not necessarily be financial, even though it is certainly welcome, a public policy at the federal level in the United States may simply incarnate as an automated transit research/testing program under the federal transit administration (FTA). The current environment is fairly conducive to the development of such program. For example, FTA has recently gained the authority in safety oversight, which brought it to par with Federal Railroad Administration (FRA), who is the regulator of railroad safety. Utilizing its newly acquired safety oversight, FTA has the potential to mandate design standard, such as those developed by the ASCE APM standard committee or other organizations. FTA may also explore the safety improvement brought by automated transit applications and explore the possibility for further implementation of automated transit technologies.

FTA, the successor of UMTA, has been largely absent from the automated transit development except the minimum data collection via National Transit Database (NTD) from those relevant transit operations. A recent attempt by FTA to discuss cooperation on PRT concepts with Swedish Transport Agency brought new hope for renewed interest or support (Federal Transit Administration, 2011). In January 2011, the US Department of Transportation (USDOT) hosted a delegation from Swedish Ministry of Enterprise, Energy and Communications, and the Swedish Transport Administration to discuss cooperation in the areas of high speed rail, livability and sustainability, PRT, and road safety. FTA conducted a break-out session on PRT. During the discussion, the United States and Sweden expressed their intent to share studies and plans, convening a workshop with key stakeholders, and conducting a technical scan of PRT practices, among other cooperative efforts. It has been several years since the signing of memorandum of understanding but no further progress is observed.

When there is a national policy and/or program that recognizes automated transit technology or applications as one of the viable alternatives among an array of multimodal transportation scenarios, local government or planning agencies will have the ability to include it in the alternative analysis and comparison, which may have the potential for an automated transit application to be selected as the locally preferred alternatives (LPA) via a lengthy EIS study process. If and only when the sequence of those events described above take place, technology suppliers may grow and prosper or large, self-funded companies may enter the supplier market, which is the pre-requisite for many transit agencies to commit public fund in acquiring ATS.

As documented in Chapter 2, there were quite a few demonstration projects and feasibility studies of AGT under the national development policy in the 1970s. For example, a number of local government agencies in Denver, CO; St Paul, MN; San Diego and San Jose in CA all participated in the feasibility study and submitted bid for the pilot project (Office of Technology Assessment, 1975). All three downtown people movers (DPMs) in Detroit, Jacksonville, and Miami have gone through the planning process as part of a special demonstration program in the 1970s (Sproule, 2009). Many corridor studies were conducted by different authorities and private interests, such as Port Authority of Allegheny County, Pittsburgh, and El Paso /Juarez to examine the feasibilities of automated transit applications. Given the short institutional memory, it is imperative to develop a library or archive the studies and methodologies for the continuity of automated transit development and progress.

7.2 LONG-RANGE TRANSPORTATION PLANNING

In order to be considered for federal funding, an automated transit application has to be included in the long-range transportation planning (LRTP) process in most urban areas in the United States. As a federally mandated and funded transportation policy-making organization, metropolitan planning organization (MPO) usually manages and maintains the LRTP for a given region. Authorized by the Federal-aid Highway Act of 1962, MPO channels federal funding for transportation projects and programs through the LRTP process. According to the "New Starts Planning and Project Development Process" shown in Figure 7.1, any new transit project will be evaluated via systems

TEA-21 New Starts Planning and Project Development Process

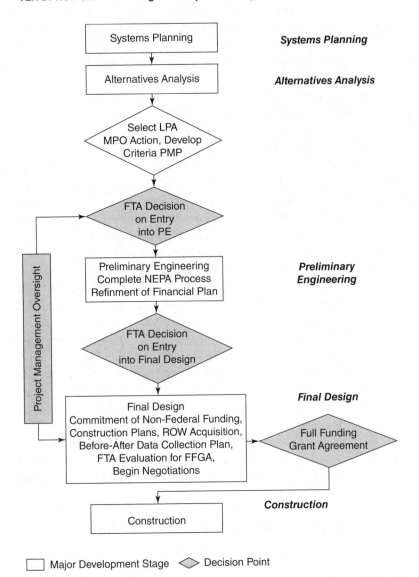

FIGURE 7.1 New Starts Planning and Project Development Process. *Source*: Federal Transit Administration, 2015. Public domain.

planning process. The winner will be selected as the Locally Preferred Alternative (LPA) before preliminary engineering and final design and construction to begin. The famous 3C process, continuing, cooperative, and comprehensive, was developed to ensure the appropriate allocation of federal funds.

Public access and participation is mandated to ensure the transparency of the LRTP process.

Instigated by the federal mandate, many MPOs utilize travel demand forecast models to evaluate the impact of various proposed transportation projects in the region (Liu, 2009). The widely accepted four-step travel demand forecast model has the ability to represent each transportation project or service in a regional network by specifying various metrics such as travel speed, capacity, and cost. The operation simulation can be performed when the estimated travel demand is distributed among the combined multi-modal transportation network in the region. Assuming national policies or public interests made it possible for automated transit applications to be included in the travel demand forecast process, the following sections explain how automated transit alternatives should be included in the evaluation process.

7.2.1 Trip Generation

The first module of four-step travel demand forecast model, Trip Generation, is to define the magnitude of total daily travel at the household and zonal level for various trip purposes (McNally, 2007). This module also explicitly translates daily activities into distinct trips and simultaneously classifies each trip into a production and an attraction. Trip generation essentially defines total travel in the region and carries the total trips into the remaining steps.

There may not be much change to the travel demand forecast model when incorporating a typical automated transit application since the basic operating characteristics of automated transit, such as DLM or driverless bus is adequately reflected in the transportation supply systems. However, significant changes may occur when PRT and/or automated personal transit (APT), driverless vehicles owned and operated by public agencies, become reality. It is speculated that more trips may be generated when driving is no longer a necessity. For example, seniors, children, or disabled, who were not able to drive before, will have the option to use automated vehicles, PRT, or APT. It is certain that more trips will be generated by the virtue inclusion of a large segment of population but the precise magnitude of the increase may be difficult to estimate as such demand is usually derived from the combination of travel time, cost, comfort and convenience, safety and security, and reliability of the transportation systems.

In the conventional travel demand forecast model, travel time and cost are usually the independent variables included in disutility of travel mode or alternatives while it is a struggle to incorporate comfort and convenience, safety and security, and reliability metrics into the econometric model as those factors may be intangible, hard to measure, or simply subjective. When automated transit applications, especially automated vehicle or APT enter the market place, the very concept of disutility associated with travel time or value of time maybe overturned.

For example, all personal travel times in today's society are treated as unproductive or disutility since time spending on driving or riding transit is not productive or compensated. The conventional model even insert a penalty on to out-of-vehicle travel time (OVTT), such as walking or waiting, as those types of travel times are usually more onerous than in-vehicle travel time (IVTT), therefore; the disutility may be magnified two to three times when a traveler makes his or her travel decisions. It is widely anticipated that, when and if APT or fully automated vehicles become reality, travelers are not required to drive the vehicle but freed to do whatever task they chose. It is possible for travelers to work, rest, or engage in leisure activities, therefore it is possible to convert travel time from disutility to positive utility, which may or may not be treated differently from working time at the office or leisure time at home. In that case, not only the magnitude of the coefficients but also the sign, positive or negative, of those coefficients may be different from today's practice.

7.2.2 Trip Distribution or Destination Choice Module

The second module of four-step travel demand forecast model, Trip Distribution, is to recombine trip ends from trip generation into trips, although typically defined as production–attraction pairs and not as origin–destination pairs (McNally, 2007). The trip distribution model is essentially a destination choice model and generates a trip matrix or trip table for each trip purpose utilized in the trip generation model as a function of activity system attributes and network attributes.

In the traditional four-step travel demand forecast mode, the distribution stage is typically formulated as either a gravity type model, or a logit type destination choice (DC) model (Liu, 2009). The logit-based destination choice model can be constructed with composite time and cost utilities reflecting accessibilities provided by all modes of travel by either using a generalized costs formulation, or the "logsum" of the mode choice model. An advantage of this approach would be to provide an opportunity for including additional and perhaps specialized attraction or destination variables, such as beaches, parks, regional malls, sports facilities, which may be served by automated transit applications. These attractions, size variables, can be made an integral part of utility calculation in the DC model.

Similarly, when automated vehicles or APT becomes a reality, it is expected that longer trips will be observed, as human driver is no longer necessary and the in-vehicle travel time may be utilized productively. The behavior may reflect a known phenomenon of "positive utility of travel" when travel time/distance is intentionally not minimized within certain limits. This is a serious challenge for urban planners who are concerned about urban sprawl, air pollution, and energy consumption when automated vehicles become commercially available or enter mass market. Another way to address such change is to reflect the re-distribution of various trip destinations via land use and social economic components that are the basis of four-step travel demand forecast model.

7.2.3 Mode and Occupancy Choice Module

Mode choice module, the third step in a traditional four-step travel demand forecast mode, effectively factors the trip tables from trip distribution to produce mode-specific trip tables (McNally, 2007). These models are now almost exclusively disaggregating models that estimate separate choice-based samples and reflect the choice probabilities of individual trip makers. Many recent mode choice models reflect current policies such as carpooling choices, taxi services, and tolls on automobiles. The most common technique used for mode choice calibration is the nested logit model. Various nesting structures corresponding to the sub-modes or modal groups presented in Figure 1.3 can be developed and mode share, or percentage of users, will be estimated for each sub-mode. These mode choice models can reflect a range of performance variables and trip-maker characteristics to produce disaggregate results that may then be aggregated to the zonal level prior to route choice (Ortuzar and Willumsen, 2001).

In order to incorporate automated transit applications in the mode choice module, an important question is: are the travel sensitivities, utility coefficients, associated with automated transit the same as with conventional transit? Given the existing automated transit applications and their operation characteristics, the answer may be a straight forward "yes." However, when automated vehicles or APT fleets become available, the mode choice model should reflect a somewhat different set of preferences compared to conventional transit or private vehicles that operate on the roads today. The comfort and convenience factor may influence passengers mode choices more than in-vehicle travel time, which becomes less relevant when traveler is liberated from the driving task, especially under congested conditions. The co-efficient for "willingness to pay" may be higher or lower depending on specific activities, so-called situational value of time, for example.

The actual differences between sensitivities in conventional transit or automated transit, especially APT, need to be estimated and validated once both types of services become reality. Both the sensitivities of various metrics and the differences of the same parameter among various modes are important to determine how each traveler will select a particular mode. For example, when comparing APT with conventional transit or private vehicles, it is anticipated that the incremental values and coefficients may be very different for the following set:

- In-vehicle travel time
- Out-of-vehicle travel time
- Tolls/fares/costs
- Value of time – implied.

Depending on the resources available, the models will either need to adopt and adapt parameters from related markets represented in the automated transit or APT travel, for corresponding activities/purposes, or be able to use new parameters of the utility expressions in the mode choice model estimated specifically for automated transit applications with a full statistical model estimation process. The latter is clearly preferred, but requires the collection and development of sufficient travel behavior survey data, which will help to develop reliable matrices of travel times and costs for individual modes of automated transit and/or APT operations based on contemporary highway and transit networks and supporting zonal land activity data.

7.2.4 Trip Assignment Module

As the last step in the traditional travel demand forecast model, Trip Assignment module is capable of selecting the most efficient routes for individual trips based on a composite "utility" score, which will eventually enable the network reach an equilibrium stage. When automated vehicles or APT fleets become available, this selection and rerouting process maybe easily implemented via the central control systems.

To realize the promise of automated transit applications, long-range transportation plans must become strategic, framing, and evaluating financially realistic alternatives that can guide elected officials and the public through the difficult choices required to balance air quality and transportation concerns. With much smaller footprint and operating flexibilities, alternatives with automated transit applications will have better chance to be selected as the LPA via comprehensive evaluation. Transportation improvement programs (TIP), which often consolidates decisions made outside the MPO process, must demonstrate links to the long-range transportation plan and how projects are selected to accomplish regional objectives. Automated transit applications may appear in the TIP only when there is enough support from local residents, public officials, and transportation professionals.

As explained earlier, the process for incorporating automated transit, especially AGT applications into the existing LRTP process by various MPOs is straight forward even when the latest activity based models are utilized. It is challenging to incorporate future automated vehicles and/or APT into the existing travel demand forecast model but it is not insurmountable. The factors preventing automated transit being included in the evaluation process is far from technology or technical capability of travel demand models. The true obstacle lies with the national policy and public interest.

7.3 OPERATIONS PLANNING

Once automated transit is identified or selected as a viable alternative in the LRTP, transit agencies or operators may develop tactical operation plans from the supply side. A transit operation planning process usually includes four basic aspects (Cedar, 2007):

- Network route design
- Time table development
- Vehicle scheduling and
- Crew scheduling

It is obvious that the fourth element, Crew Scheduling, may become much simpler or even eliminated for automated transit applications. The second element, Time Table Development, will be eliminated for APT, or on demand ATS. For most of AGT applications, such as DLM or APM applications, the operations planning should not be any different from conventional transit, which can be referenced in conventional text books (Vuchic, 2007 and Cedar, 2007). If any difference, operation planning for automated transit application should become easier as less restrictions on crew availability and regulatory oversight. For example, the working hour restriction on locomotive engineers will no longer apply since no train driver is needed. The unexpected absence of crew members will have less impact on transit services as no driver or crew is expected on the AGT trains.

As described earlier, APT will only take shape when automated vehicles are owned and operated by third parties, either public or private entity. The operations planning for APT fleet will resemble today's private automobiles or taxis more than public transit. Therefore; the operations planning for automated transit applications for the near future will focus on vehicle scheduling.

One of the great advantages of automated transit application is to provide higher quality service by increasing train frequency via shortened trains without increasing driver costs. Many transit agencies with DLM or auto-mated light rail transit (ALRT) applications have been taken full advantage of vehicle scheduling to provide higher quality service via shortened headways and/or shorter trains. For example, the DLM in Denmark has opted for shorter trains in anticipating higher service quality with shorter headways without being concerned with driver availability and cost.

A recent survey of APM systems in the airport revealed how most APM systems determine their service schedules (Gambla and Liu, 2012). Among the 14 Airport APM applications that responded to the survey, about 38% APM applications are controlled based on headways and the schedules have been shown to public. About 23% systems have fixed schedules based on departure times and schedule is also published. The rest 39% airport APM applications have their own methods to adjust the headways.

As documented in Table 7.1, SkyLink, the APM application at Dallas–Fort Worth International Airport (DFW), maintains a round trip time of 20 minutes and keeps trains de-bunched by manipulating dwell times and train speeds. The number of trains for each loop is determined via passenger head counts conducted by DFW Skylink staff. Different from those systems mentioned earlier, SkyLink does not publish any "schedule" to the public. Similarly, "AirTrain," the APM system in San Francisco International Airport deter-mines its schedule by demands. It uses passenger count and airline forecast to estimate the demands for APM services, and then develop schedules for its "AirTrain" operations.

The APM application in Pearson International Airport in Toronto, LINK, has continuous operation in single or dual train shuttle modes. Single and dual shuttle train modes have pre-defined round trip times. The central control system for LINK synchronizes the dual train mode operation, so certain headways can be maintained and adjusted according to the ridership patterns.

The foreseeable complication in operations planning for automated transit applications lies in the process of full PRT applications. There are quite a few studies that focused on PRT network design (Ma and Schneider, 1991), capacity analysis (Lowson, 2003), and empty vehicle management to reduce the passenger's waiting time (Lees-Miller et al., 2010). However, the opti-mization or dispatch of a large number of small vehicles or podcars over an inter-connected guideway network in real operations remains a challenge or to be proven.

Two researchers from Purdue University (Zheng and Peeta, 2014) have explored an approach that focuses on a theoretical mechanism on how to reduce the Guideway Network (GN) length effectively for a PRT network design. Using a solution algorithm based upon Lagrangian relaxation, the authors have developed a multi-commodity flow formulation for the GN

TABLE 7.1   Operating Schedule Management

ID	Abbreviation	Operating Schedule Management
1	DFW	Our system is computer controlled to maintain a round trip time of 20 minutes and to keep trains de-bunched. It accomplishes this by manipulating dwell times and train speeds. The number of trains on each loop is determined via passenger head counts conducted by DFW Skylink staff. We do not publish any "schedule" to the public.
2	MSP - concourse	The system is controlled based on headway schedule and "timetable" is presented to the public
3	MSP - hub	The system is controlled based on headway schedule and "timetable" is presented to the public
4	ORD	The system is controlled based on headway schedule and "timetable" is presented to the public
5	DTW	Basic Daily Operation: Both trains operate continuously (5:00 am – 10:30 pm) One train operates continuously (10:30 pm–11:30 pm) Out of service (11:30 pm–5 am) There are times (e.g., flight delays) where we do operate the express tram during this period, but not often (approx. 5–10 times/year) The above information is presented to the public via tram directory signs located at the airport
6	YYZ	Continues operation in single or dual shuttle train modes. Train control system synchronizes the dual train mode operation. Single and dual shuttle train modes have pre-defined round trip times.
7	SFO	Passenger count system and airline forecast
8	MEX	The system is controlled based on headway schedule and "timetable" is presented to the public
9	TPA garage monorail	The 5-car system is operated in a continuous pinched loop mode with three trains removed from service during a 5-hour maintenance window each morning except for 1 day each week when the system is shut down for 5 hours for track maintenance.
10	TPA APM shuttle	Each leg is operated in two-train operation in lock synchronous mode with one train per leg removed during a 4-hour maintenance window each morning.
11	DIA	The system is controlled based on headway schedule and "timetable" is presented to the public

Source: Gambla and Liu, 2012.

network design and presented a case study to demonstrate the effectiveness of the proposed methodology.

Similarly, some studies (International Transport Forum, 2015 and Fagnant and Kockelman, 2014) have anticipated the advent of shared autonomous vehicle (SAV) operations or APT as defined in earlier chapters of this book. Simulating the operations of SAV; researchers have estimated the need for the number of vehicles, potential for increased vehicle miles travelled (VMT), and reduction in environmental impact. However, the actual dispatch or operation implementations of such applications still remain to be seen.

Another school of thought is that the rapid development of level 3 and level 4 automated cars and buses may eclipse the idea of PRT and conventional LRT by enabling automated operation of cars and buses on public rights of way and in mixed traffic. The counter argument is that PRT, with dedicated track/right of way, will still have advantages over Automated Vehicle that operates in mixed traffic, prone to congestion and delay especially when more increased VMT or passenger miles travelled (PMT) are expected.

REFERENCES

Anderson, J. 2009. "How to design a PRT guideway." In: Automated People Movers 2009: Connecting People, Connecting Places, Connecting Modes. Proceedings of the 12th International Conference, Atlanta, GA, 2009

Carnegie, J., and P. Hoffman. 2007. "Viability of Personal Rapid Transit in New Jersey." Prepared for New Jersey Department of Transportation by Alan M. Voohees Transportation Center, Rutgers, The state University of New Jersey.

Cedar, A. 2007. *Public Transit Planning and Operation: Theory, Modeling and Practice*, Elsevier.

Committee of Automated People Mover Standards. 2006. "Automated people mover standards." Available at http://www.asce.org/templates/publications-book-detail.aspx?id=7978. Accessed in August 2015.

El-Aasar, M., R. Willard, and C. Hibbs. 2006. "Automated small vehicle transit system structural and architectural research study for a university campus." Report No. KS-06-5.

Fagnant, D., and K. M. Kockelman. 2014. "The travel and environmental implication of shared autonomous vehicles using agent-based model scenarios." In: Proceedings of the 93rd Transportation Research Board Annual Conference, Washington, DC, January 2014.

Federal Transit Administration. 2011. "Examples of recent technology transfer programs." Available at http://www.fta.dot.gov/printer_friendly/about_FTA_9361.html. Accessed in August 2015.

Furman, B., et al. 2014. "Automated Transit Networks (ATN): A review of the state of the industry and prospects for the future." Mineta Transportation Institute. MTI Report 12–31.

Gambla, C., and R. Liu. 2012. "Airport Cooperative Research Program (ACRP) Report 37 A: Guidebook for Measuring Performance of Automated People Mover Systems at Airports." Sponsored by the Federal Aviation Administration (FAA). Transportation Research Board (TRB) of National Academies. Washington, DC, 2012.

International Transport Forum. 2015. "Urban mobility system upgrade: how shared self-driving cars could change city traffic." Corporate Partnership Board Report. 2015.

Lees-Miller, J., et al. 2010. "Theoretical maximum capacity as benchmark for empty vehicle redistribution in personal rapid transit." *Transportation Research Record: Journal of the Transportation Research Board*, vol. 2146, pp. 76–83.

Liu, R. "Weekend travel demand and mode choice models–Final report." Prepared for and submitted to Federal Highway Administration (FHWA), New Jersey Department of Transportation, June 2009.

Liu, R. 2013. "Tri-Annual Strategic Plan for AP040: Automated Transit Systems (ATS) Committee." Available at https://sites.google.com/site/trbcommitteeap040/files. Accessed in December 2014.

Lowson, M. 2003. "Service effectiveness of PRT vs. collective corridor transport." *Journal of Advanced Transportation*, vol. 37, no. 3, pp 231–241.

Ma, J., and J. Schneider. 1991. "Designing Personal Rapid Transit (PRT) networks." *Journal of Advanced Transportation*, vol. 25, no. 3, pp 247–268.

McNally, M. 2007. "The four step model." In: *Handbook of Transport Modeling*, edited by Hensher and Button, Pergamum, 2007.

Office of Technology Assessment. 1975. "Automated guideway transit: an assessment of PRT and other new systems." Prepared at the request of the Senate Committee on Appropriations Transportation Subcommittee, June 1975. NTIS order No. PB-244854.

Ortuzar, J., and L. Willumson. ARRB Group Limited. 2001. "Modeling Transport." Monograph.

Page, T. 2012. "Automated Transit Network Feasibility Evaluation." San Jose Mineta International Airport, San Jose, CA. Aerospace Report No. ATR-2012 (5629)-1 REV A.

Sproule, W. 2009. "Somewhere in time–a history of automated people movers." In: Automated People Movers 2009: Connecting People, Connecting Places, Connecting Modes. Proceedings of The 12th International Conference, Atlanta, GA.

UITP. 2013. "Automated Metros in the world." Available at http://metroautomation.org/. Accessed in December 2013.

Vuchic, V. 2007. *Urban Transit Systems and Technology*. John Wiley & Sons, 2007.

Zheng, H., and S. Peeta. 2014. "Design of personal rapid transit networks for transit oriented development cities." NEXTRAN Project No. 081PY04.

CHAPTER 8

BUSINESS MODELS FOR AUTOMATED TRANSIT APPLICATIONS

There is a variety of business models when it comes to the operation and management of automated transit applications. This chapter provides a systematic review of those diverse practices, which may be referenced by different entities when contemplating with automated transit applications. As concluded in a recent study (Carnegie and Hoffman, 2007), automated guideway transit (AGT), especially personal rapid transit (PRT) possesses the virtue of sustainability owing to its small footprint, lower cost, and lower impact on the environment. On the other hand, their small sizes and low-key profiles have fostered a number of applications around the world without garnering any major headlines.

The perfect safety record, no casualty whatsoever, during the past half century of operations has also shielded AGT, particularly group rapid transit (GRT) from negative headlines. However, their potential as unique solutions to urban circulation and congestion problems should not be suppressed any longer. Automated transit, especially AGT and PRT, applications deserve the proper highlight in the array of modern technology development, public welfare, and successful business models. Since the spectrum of automated transit technology stems from various geographic, operational, and institutional settings, several business models should be developed and promoted to foster further development of the technology and to expedite the application processes (Liu et al., 2009).

Automated Transit: Planning, Operations, and Applications, First Edition. Rongfang (Rachel) Liu.
Copyright © 2017 by The Institute of Electrical and Electronic Engineers, Inc. Published 2017 by John Wiley & Sons, Inc.

It is observed that recent driverless metro (DLM) or medium capacity automated people mover (APM) applications have largely followed the conventional transit business model, funded by either a sole or combination of public entities. Meantime, the search for a business model of PRT or automated transit network (ATN), small vehicle or podcar applications, may very well extend into uncharted territory. Furman et al. (2014) suggested that the most viable and promising business model for PRT appears to be a real estate investment approach. His reasons include that private funds will be invested more readily in an ATN project with the expectation of significant future incomes from the appreciation of urban land and building values at or near stations. In reality, it is observed that both private and public funds have been invested on PRT or ATN applications.

The implementation of an automated transit application, such as AGT, PRT, or ATN, like any large urban infrastructure project, is not a simple consumer choice, such as an individual deciding to purchase a new computer. Designing and procuring an automated transit application is actually an intermedia brokerage that procures and operates transit services, which need to be sold to individual consumers, passengers, or transit riders. The procurement process may be somewhat comparable to a large corporation acquiring a company-wide communications system that involves hardware, software, and technical staff (Furman et al., 2014). It is the type of purchase that affects day-to-day operations and ultimately the corporation's survival.

Moreover, automated transit procurements are further complicated because they are in the public realm, and require the consensus of contending forces to expend millions—if not billions—of dollars. Automated transit applications mandate detailed engineering design, environmental impact review, regulatory approvals, and system certifications. Safe and reliable operations of the automated transit services are necessary but not sufficient to warrant the success of the automated transit application since the other side of the equation is ridership and customer satisfaction, which are delicately balanced between supply and demand as well as the combination of art and science.

A business model establishes the core operational requirements essential to implement AGT applications safely and efficiently. It also addresses the spatial, temporal, and institutional structures associated with implementation of AGT technology. A business model highlights strategies to deal with incremental risks and liabilities associated with new or expanded operations of AGT systems. It also identifies potential funding sources, market-analysis procedures, and ridership shares when compared with other modes based on domestic and international experience. Several business mode structures are documented and explained in the following sections.

8.1 PUBLIC OWNER AND OPERATOR

The most likely public owner of an automated transit application is a local transit agency even though there are wide range of variations in customer markets, funding mechanisms, performance measures, and organizational structures. Transit agency governance has an influence over an agency's ability to adopt new technologies and continuous operations of respective technology applications. According to Booz Allen Hamilton (2011), there are five most common governance structures in regional organization models for public transportation:

1. State-wide transit agencies: Agencies owned and operated by a state, such as the Maryland Transit Administration (MTA) and New Jersey Transit (NJ TRANSIT).
2. General purpose authorities: Transit organizations formed where state law permits the establishment of a transit agency outside of local government. General purpose authorities are usually established by state-enabling legislation initiated by local actions and support. Examples include Ohio's transit authorities and Florida's transit districts.
3. Special purpose authorities: Transit agencies created as a result of a specific act of the state legislature, such as the Bay Area Rapid Transit system, or the Utah Transit Authority (UTA).
4. Municipal agencies: Transit agencies operated by existing local governments, such as Charlotte Area Transit, King County Metro, or the San Francisco Municipal Transit Agency.
5. Joint powers authorities: Special local arrangements established to jointly exercise powers, such as the Virginia Railway Express and the Trinity Railway Express in Dallas-Fort Worth.

There are several funding alternatives available to initiate and/or support the operation of automated transit applications (Gannet Fleming, 2014). In the United States, most transit applications, such as Morgantown GRT and the three downtown people movers (DPM) are funded by the federal grants and operated by local transit agencies. Morgantown GRT receives payments directly from the transportation fees included in each West Virginia University student and the payment makes up the majority of fares, about 95%, collected for the system (Hsiung and Stearns, 1980). Miami Metromover has never charged a fare for its riders, Jacksonville Skyway was made free since February 2012, and Detroit DPM charges a symbolic 75 cents for the ride.

Therefore; the farebox recovery ratio (FRR) for those DPMs is either zero or extremely small.

FRR measures the fraction of operating expenses that are met by the fares paid by passengers. It is computed by dividing the system's total fare revenue by its total operating expenses. Similar to all other transit operations, the fare box recovery rates for all automated transit applications are very low, less than 30%, which only cover a small percentage of their operating expenses, therefore; they all need government subsidy. A recent study (Gannet Fleming, 2010) has discussed various funding alternatives for the Morgantown GRT applications based on how transit projects are most often funded. It is noted that the likelihood that all of the necessary funding for the Morgantown GRT improvement will come from one source is remote. Federal sources have a matching component and other sources noted will most likely only generate a portion of what is necessary to complete the task. Therefore, each entity contemplating automated transit applications should consider most of these alternatives or evaluate them in combination when determining the best funding option for both capital and operation and maintenance expenses.

The public owner and operator business model for automated transit application are also observed in international locations. For example, many DLMs recently implemented in Europe and Asia belong to the intermediate capacity line-haul category. The use of automation permits shorter trains which can provide more frequent service, especially during the off-peak hours. Such line-haul concepts often replace aging long subway or transit trains with more frequent services without increasing operating costs. The operator/owner relationship for converted DLM is largely inherited from the previously conventional manually operated transit lines, therefore; most of them are publicly owned and operated applications.

As documented in previous chapters, Paris Metro Line No. 1 was converted to fully automated services without major interruption to passenger traffic. As part of the Paris Metro, both DLM Lines, Line No. 1 and Line No. 14, are operated by the Regie Autonome des Transports Parisan (RATP), a public transport authority. Similar to many transit applications in America, Paris Metro obtained the major shares of capital investment from the French Government. It seems that funding from the government will continue to be the main source of automated transit applications in France, especially Paris Metro. For example, in June 2014, the Board of Ile-de-France Transport Authority has approved 100 million Euros to the automation of Paris Metro Line 4 (Barrow, 2014).

Another large category of publicly owned and operated automated transit applications are those airport APM systems. As noted by Neufville (1999), many peripheral services in the US airports, such as concessions in the terminal and the rental car businesses, are privatized but the major aspects of airport operation must remain under government control, because of the public interest in those facilities. The public interest and government control become even more prudent after the 9/11 terrorist attack in 2001. The majority of commercial airports in the United States are owned and operated by Port Authority, a public entity formed by the local or state government agencies.

Most airport APM applications are publicly owned and financed in the United States. The potential funding source may come from an individual program or combination from the following categories:

- Airport improvement program (AIP)
- Passenger facility charges (PFC)
- Airport generated revenues
- Airport revenue bonds

The AIP, administrated by Federal Aviation Administration (FAA), provides grant assistance to public use airports for capital improvements that enhance safety, capacity, security, or the environment. Many airport APM applications shuttling among different terminals and/or connecting parking, terminal, and other aviation facilities are designed and constructed to improve safety and security conditions, expand capacity and/or minimize environment impact. They are often great candidates for AIP grants.

As part of a federal program administered by FAA, PFC allows the collection of fees up to $4.5 for every enplaned passenger at commercial airports controlled by public agencies. It is straight forward to utilize PFC when funding airside APM applications but there might be limitations in funding landside APM in the airport, especially if the application extend off airport property (Lea+Elliott, 2010).

There are different sources for an airport in the United States to generate revenue. For example, airline landing fees, vehicle parking charge, surcharges applied to car rental and terminal concessions are typical and widely known airport revenue streams. An airport may also generate revenue via off-airport commercial vehicle access fees, customer facility charges (CFC) from users of rental car facilities. As part of the reinvestment of airport generated revenue, those funding sources may be used to construct and/or operate airport APM applications.

Another common form of financing for airport infrastructure, including APM applications, is the proceeds from the sale of revenue bonds. Other types of bonds, such as general obligation (GO) bonds backed by local tax revenues, special facility bonds backed by commitments from facility users, and commercial papers, are less common but not unheard of in financing airport APM applications.

Breaking away from the total reliance on government subsidy for transit services, a few transit agencies, such as Hong Kong Regional Transit Authority (RTA), have successfully covered their operating expenses using revenues from combination of transit operations and real estate development (Padukone, 2013). Similarly, two European examples of the real estate-based approach to transit development are in London and Copenhagen (Furman et al., 2014). The London Docklands Development Authority in the 1980s invested in a relatively low-cost automated light rail transit (ALRT) application that catalyzed large-scale investment in office towers. The Dockland ALRT has been upgraded and expanded in several stages, funded by revenues from real estate development. In Copenhagen, a special development authority was created to develop a large tract of land between the city center and the airport. The authority selected DLM application to provide high levels of transit services to the development. The DLM application was largely funded by anticipated increases in land value.

8.2   PRIVATE OWNER AND OPERATOR

A continuous monitoring effort on APM development documented that about one-third of APM applications are located in private institutional settings (Liu and Huang, 2011). As shown in Figure 8.1, about one-third of the airport APM applications are operated by the airport owners, another one-third are contracted to Bombardier, a main APM developer and technology supplier, and the remaining one-third have been spread among several smaller companies. Typical examples of privately owned and operated automated

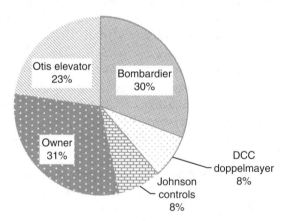

FIGURE 8.1   Owner and Operator Split for Airport APM Applications. *Source*: Liu and Huang, 2011.

transit applications include the APM in Walt Disney World, Monorail in Las Vegas, and Ultra PRT in Heathrow International Airport in London.

Inspired by his early experience in the South Park Demonstration Project, Mr. Disney has implemented many APM applications in his amusement parks. Those APM applications vary by length, size of the cars, speed, and sometimes they are disguised by various dramatic surroundings to make the ride more exciting. Figure 8.2 showcases a few examples of APM applications in Walt Disney amusement parks.

Another privately owned and operated automated transit application, Las Vegas Monorail (LVM), actually provides a public service. It connects the unincorporated communities of Paradise and Winchester (Las Vegas Monorail, 2011) even though most of its riders come from the Las Vegas Strip. Registered as a not-for-profit corporation in the State of Nevada, Las Vegas Monorail Company funded the construction of LVM even the State of Nevada provided assistance in bonding process, no public money was used as the capital investment.

The LVM had a rough opening or commissioning period due to various parts falling off the elevated guideways. During its operation for the past decades, LVM has experienced occasional shut downs due to various electrical or mechanical issues. It also handled large crowds, such as 2005 Consumer Electronics Show (CES). The LVM generates revenue from ticketed passengers and from corporate sponsors.

To support its ongoing operation, LVM has debuted various sponsorship programs, such as the red "Money Train" by Bank West and "Star Trek" themed trains by Paramount Studios (Anderson LLP, 1999). Branding rights for the seven stations and the nine trains of LVM are available for sponsorship, which often fetch multiple millions of dollars for each sponsorship. For instance, Hansen Beverage sponsored a monorail train, featuring its Monster Energy drink. Nextel Communications created a totally themed pavilion by branding the largest station, adjacent to the Las Vegas Convention Center. Since the Sprint–Nextel Merger in late 2005, Nextel Central has been rebranded as Sprint Central.

The third example of privately owned and operated automated transit application is the newly inaugurated Ultra PRT system in Heathrow Airport in London. Comparing to the mostly publicly owned and operated airports in the United States and other international locations, UK has privatized its airports in a steady pace and handed over about half of its commercial airports to private enterprises (Calder, 2014). As the new incarnation of British Airport Authority (BAA), Heathrow Airport Holdings Limited is the UK-based operator of London Heathrow Airport, which commissioned the Ultra PRT application in 2011 to connect a car park with the main terminal.

(a)

(b)

FIGURE 8.2   (a, b) Examples of APM Applications in Disney World. *Source*: Pixabay. Creative Commons CC0.

Ultra Global, Inc. designed and operates the PRT application in Heathrow International Airport. Prior to the selection of Ultra PRT technology, a tender process was initiated when it was determined that more than one supplier could provide the technology. About 30 initial expressions of interest were received and evaluated. The list was shortened to four, based on the following criteria:

- The supplier must have a test track carrying passengers or being built;
- The system footprint must meet the dimensional requirements of the existing tunnel at Heathrow International Airport;
- The vehicle should be able to carry four passengers plus luggage and meet additional requirements.

Assembling a team of experts, primarily external, in topics such as software development, communications, transportation, and simulation, BAA evaluated detailed responses to a wide array of technical issues, price quotes for the first phase, and intention for the entire airport build-out plan. Ultra Global, Inc. was awarded the contract. The vehicle, communication and control systems, and the civic infrastructure were included in one contract. In financial terms, the contract was open-ended, with monthly payments.

Being extremely experienced with understanding passengers, BAA set passenger perception as their top priority. BAA also took control of delivery of infrastructure items, about 70% of the project, even the authority tried to not micromanage the Ultra application construction and operations. BAA has invested in Ultra directly, which may be perceived as conflicting interests by some but most accept that it was necessary to complete the project. The direct involvements from BAA in certain items that had not been demonstrated previously helped mediate or reduce risk exposure for Ultra, which was adaptive and agile in technical development.

Looking back, David Holdcroft, the project manager for BAA on Ultra application indicated that there were few technical issues with Ultra PRT. A few problems tended to be caused by communication and organizational shortcomings. For other entities that contemplate PRT applications, Mr. Holdcroft recommends project managers and transit planners:

- Require extensive testing and mockups from suppliers;
- Conduct extensive simulations and emulations;
- When breaking new ground, condition payments to milestones;
- Conduct a thorough assessment of organizational ability of the proposers.

Similar to other private owner/operator of automated transit applications, Ultra PRT in London Heathrow Airport also made the most of sponsorship opportunities. Located adjacent to the T5 Business Car Park in Heathrow International Airport, the Thistle Hotel approached BAA for access to the Ultra PRT for their hotel guests. As the result, Thistle Hotel provided funds for the construction of an access gate to PRT station directly from the hotel. In turn, the Thistle Hotel charges its guest 5 pounds per passenger per direction, the majority of which is passed to Heathrow. The latest statistics show that each month about 2000 hotel guests use the Ultra PRT. Direct access to the "podcar" systems to the Terminal 5 in Heathrow Airport has increased the value of the Thistle Hotel.

Another sponsor of the Heathrow "podcar" is Marriott International Hotel Groups, which spent a six figure sum to promote its hotel brands. The sponsorship includes advertising on vehicle interior/exterior, station areas, and depot walls. The advertising messages were visible to almost half of all passengers that travel to Terminal 5 by road (Ultra Global Inc., 2014).

As noted by several studies (Office of Technology Assessment, 1975; Furman et al., 2014), other potential markets for AGT applications, especially PRT application, may include the use of simple shuttle and/or loop applications as horizontal elevators for airports, shopping centers, remote parking areas, hospitals, and other similar locations. There is evidence that such application may be financially viable without federal assistance because of the expanded architectural freedom, increased access to facilities, and/or improved land use values made possible by the AGT applications.

8.3 PUBLIC AND PRIVATE PARTNERS

On a positive note, the last two sections demonstrated both public and private entities may succeed in owning and operating automated transit applications. Several studies have shown that PRT schemes should be able to recover their full capital and operating costs through the fare box. However, the initial capital outlay required is expected to be more expensive than for an equivalent bus scheme. Many local authorities with limited budgets will therefore be tempted to go for the apparently cheaper and better understood bus alternative rather than the riskier and unproven solution of PRT or automated transit all together. For these reasons, most proponents of PRT believe that procurement through a private finance initiative (PFI) is to be preferred. This will enable involvement by the private sector who can be expected to accept some of the risk, both technical and financial, in return for a share of the profits.

Automated transit applications are of such scale and complexity that they require professional project management with contingencies for unforeseen circumstances. An automated transit implementation can be structured in several ways. A public or private entity can procure the various components and then assume responsibility for assembly and integration into a working system. Most governments and corporations do not have and do not want to acquire such technical expertise and project management skills. For them, there are alternative strategies known as design-build (DB), design-build-operate (DBO), and design-build-operate-transfer (DBOT) in which the automated transit supplier and builder integrate and then turn the project over to the owner after an agreed number of years.

On the other hand, federal and local government funds for transit projects are constrained by larger budgetary and debt concerns. The formation of public–private partnerships (PPP) is one way to introduce other sources of funding for urban projects. In theory, this could include automated transit technology suppliers and constructors. Many states are enabled to let private parties submit unsolicited proposals for PPPs to solve transportation issues. For example, regarding transit projects, the Regional Transportation District in Colorado has been particularly successful in implementing PPP projects in the Denver metropolitan area 56 (Daniel et al., 2014). Each state and/or a development agency within a state typically have developed their own unique requirements for PPPs. If transit industry has no interest or is excluded from the process, private investors, working alone, will have great difficulties to implement an automated transit project in an urban area, even if it looks profitable. Private investors must identify a return on investment (ROI) that is unlikely to arise from fare revenues alone.

As presented earlier, there is potential profit from increases in land value that an automated transit application will create. This is how Hong Kong's metro, London's Docklands Light Railway, and the DLM in Copenhagen were financed. Moreover, solar-equipped automated transit applications may generate power beyond its own use and excessive electricity can be sold in urban areas, creating another revenue source. Other revenue sources may come from selling the use of conduits built into the guideway network to house power and communication wires and cables.

As one of the emerging PPP format, government-industry consortia are widely used throughout Europe and Asia as a means to conduct research and to penetrate the commercial market. Recent PRT application in Masdar City, Abu Dhabi helps demonstrate the PPP approach for automated transit applications. The investment branch of the Abu Dhabi government has initiated the Masdar City PRT project before 2007. The main function of the Masdar City PRT is to provide internal circulation for a zero emission township. Planned by Foster+Partners and engineered by Mott MacDonald, the large district is based on traditional desert city forms of low-rise, high-density settlement with pedestrian circulations. Private automobiles are banned from entering, with circulation primarily by pedestrians and by PRT, which will eventually be augmented by a metro. The business model for the Masdar City PRT may be labeled PPP as it involves government branch working with private developments.

Similarly, the process implemented in Suncheon Bay, South Korea is often labeled as DBOM. It is understood that the Suncheon Bay PRT was designed, built, and operated by Vectus, the technology supplier. The ownership will be transferred to the city after 30 years (2getthere, 2011).

In addition, several models for PFIs are possible, known as DBFO (Design, Build, Finance, and Operate) in some countries and BOT (Build, Operate, and Transfer) in others. In all cases they involve establishing a formal legal "consortium" agreement between the public and private sector partners involved to clarify their respective roles, responsibilities, and liabilities. The funding might come from industry and banks, and also from the developers of new residential and retail parks, and out of town superstores and other leisure facilities, who should be keen to facilitate, if not actually provide, public transport links that will connect existing population centers to their new sites.

The PPP arrangement has several advantages. The best talent of industry specialties can be concentrated on a particular development site. Scarce resources, including personnel; capital; and facilities, can be conserved by avoiding competition among participants. Government expenditures are reduced through cost sharing with industry. Because the government is a participant, there is mutual interest in commercialization of the product. Both the government and industry stand to get a return on the initial investment. To strengthen the price advantage of the consortium in an initial international competition, the government can waive the recovery of cost provisions for the industry participants.

These advantages, available to foreign AGT system developers, have gradually found its places in the United States as alternative arrangements that offer institutions opportunities to improve the efficiency of transit research and development and to accelerate the rate of transit innovation and improvement. There are quite a few stakeholders in the process of automated transit implementation process and many different arrangements may work for a particular location.

Similar to the applications of light rail transit (LRT) or buses, automated transit applications are typically procured by a local authority or a public transport operator, and may be financed through a PFI. The main stakeholders in this type of PPP structure may include:

- **Local authority:** the planning authority and owner of the infrastructure on which the scheme will run;
- **Operating company:** the operator of the AGT application;

- **Technology supplier and system integrator:** provider for the vehicles, control center, and communications systems;
- **Infrastructure supplier:** a contractor to implement the necessary civil engineering facilities, including the guideway and stops, and buildings for the control center and depot;
- **Managing consultant:** project manager to oversee the overall implementation and ensure co-ordination between the technology and infrastructure suppliers;
- **Government:** agencies for certification and perhaps funding.

Other funding partners may include banks and the developers of the sites directly and indirectly served by the AGT application. Other groups that should be consulted will include:

- neighboring local and regional authorities;
- emergency services;
- local community;
- passenger interest groups;
- special needs groups;
- media.

As suggested by Fishelson et al. (2013), automated transit applications especially those isolated, fully automated systems are developed from the ground up in areas with the greatest need and opportunity. Over time, such systems can expand and become interconnected. This is an alternative to the integrated deployment strategies developed by the automated highway systems (AHS) program, which had limited success in the United States.

On the other hand, vehicle automation has never settled on a specific deployment strategy because of disagreements among many of its members. However, much of the research did include an integrated approach, whereby the technology progresses over time from manual driving to full automation. Once the fully automated vehicles become available, the business model for automated transit, especially APT, may take on an entirely different shape.

The newly established transportation network companies (TNC), such as Uber, Lyft, or Ride Scott, may be well adapted to serve as transportation service providers. If the vehicle is fully automated, the operating cost of taxi or automated taxi will be significantly reduced. Given the limited time period, each private vehicle is actually utilized by its owner and enormous cost associated with owning a private vehicle, such as purchasing the vehicle, paying for insurance and maintaining the vehicle, obtaining space to park or maintaining the vehicle, many people may simply chose to summon an automated taxi or APT when they need to travel.

It is likely that Uber or Lyft will continue its operation using a fleet of automated vehicles when it becomes available. On the other hand, it is also possible for a public transit authority to provide personal transportation service using a fleet of automated vehicles, thus a new mode, automated personal transit (APT) may take shape: the vehicle is automated and no need for driver, the service is "personal" since each passenger or traveling party may have a dedicated vehicle transporting them from their origin to their destination. The service is "public," not "private" since the automated vehicle will be operated by a third party or public entity. The service is still considered "transit" as any traveler may purchase such service and they do not need to own the vehicle or being related to the operator.

As a hybrid product of automated vehicle and PRT, the anticipated APT may become the most preferred travel choice due to its direct origin to destination services and care-free public or third party vehicle ownership. It should also be pointed out that congestion delays may continue to be experienced, or even get worse, since driverless cars will undoubtedly increase vehicle miles traveled (VMT). Therefore, the advantages of PRT with dedicated track and/or right-of-way may still be relevant.

Given the rapid development in the areas of vehicle automation and automated transit applications, the challenging question today is whether the United States should maintain the vast Eisenhower-era Interstate highway network and the auto-addicted way of life it has brought (Furman et al., 2014). The sustainability of such life style is often questioned and widely criticized. But the status quo approach will not be able to bring us out of those dead end streets. It is often agreed that the urban transportation infrastructure in the United States should be transformed into something cleaner, safer, and more economical. The quests are deflected by so many diverse directions or stonewalled by resistance to change or give up the current life style. With a poised national priority, automated transit applications with variety of capacities, speeds, and footprints can play a major role as part of a larger agenda of shifting modal balance to walking, biking, car-sharing, and mass transit.

REFERENCES

2getthere. 2011. "Rivium works." Available at http://www.2getthere.eu/rivium-works/. Accessed in October 2015.

Anderson, LLP. 1999. "MGM mirage tender offer." U.S. Security and Exchange Commission. Available at http://www.secinfo.com/dSq2u.6Uy.4.htm. Accessed in August 2015.

Barrow, K. 2014. "Funding agreed for Paris Metro Line 4 Automation." *International Railway Journal*, June 9, 2014. Available at http://www.railjournal.com/index.php/metros/construction-begins-on-grand-paris-express-metro.html?channel=525.

Booz Allen Hamilton in Association with P. N. Bay. 2011. "Regional organizational models for public transportation." Prepared under TCRP Project J-11/Task 10, Transportation Research Board, Washington, DC.

Calder, S. 2014. "What Glasgow, Aberdeen and Southampton airports sell-off means for passengers." *The Guardian*. October 17, 2014.

Carnegie, J., and P. Hoffman. 2007. "Viability of personal rapid transit in New Jersey." Prepared for New Jersey Department of Transportation Bureau of Research and NJ Transit, February 2007.

Daniel, J., H. Schachter, and R. Liu. 2014. "Feasibility and efficacy of public transportation partnerships." Submitted to Federal Highway Administration, U.S. Department of Transportation/New Jersey Department of Transportation, March 2014.

Fishelson, J., D. Frechleton, and K. Heaslip. 2013. "Evaluation of automated electric transportation deployment strategies: integrated against isolated." *IET Intelligent Transport Systems*, vol. 7, no. 3, pp. 337–344.

Furman, B., et al. 2014. "Automated Transit Networks (ATN): A review of the state of the industry and prospects for the future." Mineta Transportation Institute. MTI Report 12–31.

Gannet Fleming in Association with Lea + Elliott et al. 2010. "PRT facilities master plan." Prepared for West Virginia University, June 2010.

Hsiung, S., and M. Stearns. 1980. Phase I Morgantown people mover impact evaluation: Final report. Report No. UMTA-MA-06-0026-80-1.

Las Vegas Monorail. 2011. "Ridership." Available at http://www.lvmonorail.com/. Accessed in August 2015.

Lea + Elliott. 2010. "Guidebook for planning and implementing automated people mover systems at airports." ACRP Report 37, Transportation Research Board, Washington, DC.

Liu, R., and Z., Huang. "A composite index: performance measures of automated people mover systems at airports." *Proceedings of Transportation Research Board (TRB) 90th Annual Meeting*, January 2011, Washington DC. Peer reviewed.

Liu, R., D. O. Nelson, and A. Lu. "Business model for shared operations of freight and passenger services (09-2547)." *Transportation Research Record : Journal of Transportation Research Board*, vol. 2547, pp. 86–92. 2009. Peer reviewed.

Neufville, R. 1999. "Airport privatization issues for the United States." Massachusetts Institute of Technology, draft report.

Office of Technology Assessment. 1975. "Automated guideway transit: an assessment of PRT and other new systems." Prepared at the Request of the Senate Committee on Appropriations Transportation Subcommittee, NTIS Order No. PB-244854, June 1975.

Padukone, N. 2013. "The unique genius of Hong Kong's public transportation system." *The Atlantic*, September 10, 2013.

Ultra Global. 2014. "Heathrow performance statistics." Available at http://www.ultraglobalprt.com/wheres-it-used/heathrow-t5/. Accessed in December 2015.

CHAPTER 9

LESSONS LEARNED

Given the long and capricious development process of automated transit technologies and applications, it is essential to sort through the volumes of studies, project reports, and other research materials in order to extract accurate and reliable information for potential stakeholders. Development and deployment of the latest and effective automated transit systems has been the goal of a number of government agencies, private institutions, and individual entrepreneurs for more than half of a century. This continuous effort has been driven by the need to find a way to improve operating safety and efficiency and relieve urban congestion while reducing air pollution, minimizing dependence on oil, and reducing or eliminating the need for transit subsidies.

To improve safety and efficiency and reduce congestion, it is necessary to think outside of the conventional transit or prevalent single occupancy vehicle (SOV) boxes. Without prejudice, transportation planners, engineers, and decision makers should seek to discover transit-system characteristics that would fulfill the desired needs of modern society. Of course, in order to push or pull massive SOV drivers or private vehicle users out of their comfortable, private cocoons, any new system meeting the objectives related to safety, efficiency, and congestion has to be designed to minimize costs while maximizing ridership and providing required levels of capacity, safety and security, reliability, and comfort/convenience. Not to mention, it should also

Automated Transit: Planning, Operations, and Applications, First Edition. Rongfang (Rachel) Liu.
Copyright © 2017 by The Institute of Electrical and Electronic Engineers, Inc. Published 2017 by John Wiley & Sons, Inc.

consume minimum energy, emit less pollution, and be integrated with land use. The combined criterion certainly sounds like a very tall order. However, when integrated with land-use patterns and complemented by the intermodal transportation systems or conventional transit applications, the automated transit technology may have the potential to fill such niche.

There are a number of important lessons learned after examining the automated transit development history and diversified applications. As the year 2014 marks the 50th anniversary of the first fully automated prototype automated transit system, it is logic to ask why there were not more similar automated guideway transit (AGT) applications in the United States. Why does the personal rapid transit/automated transit network (PRT/ATN) idea resurges every decades or so? Does the concept of automated transit or automated vehicle hold much appeal? Is automated transportation a "Promised Land" or simply a mirage? In searching for answers, we have learnt quite a few lessons, if not all the answers to the quests we have set out to accomplish.

9.1 DRIVING CAN BE REPLACED

Observing their teenage daughters/sons or the entire generation of millennials, many parents have been shocked and/or puzzled about the lack of interest in personal driving by their children. Personal driving has been considered to allow more "freedom" by the baby boomer generation while millennials rather browse the internet, text via mobile phones, or immerse themselves in online games while chatting with friends via multiple channels.

What options people select is largely decided by their own belief, experience, and relationships. Since driving is closely linked to freedom for the baby boomer generation, this may have created the impression that driving is the desired task. For the millennials that grew up in the era of internet, social media, and virtue reality games, the need to meet with friends face-to-face becomes secondary. Burdened by the distance, time, and monetary expenses of travel, many millennials would prefer to work from home or telecommute, communicate with friends via social media, and/or be driven to various places when travel is not avoidable.

After occupying the central stage of American life for more than a century, driving is finally being realized by a whole generation as not a completely desirable task, and it can be replaced by machines. When the 90 minutes each day for 210 million drivers in the United States behind wheels (Langer, 2015) can be replaced by more enjoyable tasks or productivities, more and more people welcome the change or look forward to the promises of automated vehicles or automated transit services.

9.2 PUBLIC POLICY: A DOUBLE-EDGED SWORD

As documented in Chapter 2, automated transit systems, particularly AGT technologies, were inaugurated in the large historical background of "landing on the moon" and the newly established Urban Mass Transportation Administration (UMTA) of the U.S. federal government. Over the course of years and decades, governmental policies and priorities change, often dramatically. One example is that in 1976, UMTA—predecessor of today's Federal Transit Administration (FTA)—decided to abandon PRT and related AGT research programs after a congressional assessment opined there was little of value in them (Office of Technology Assessment, 1975).

Under congressional and local pressure from West Virginia University, UMTA completed the then-controversial Morgantown GRT project and several socioeconomic research programs in the late 1970s and 1980s, but the United States Department of Transportation (DOT) ceased funding new AGT projects and research. The downtown people mover (DPM) demonstration program absorbed available innovation funds, while metropolitan planning organizations (MPOs) across the nation were influenced by the congressional assessment to exclude automated modes from modal agendas of long-range transportation plans. Perhaps this is why the deliberations from MPOs are dominated by bus, bus rapid transit (BRT), and conventional rail, not any AGT or PRT alternatives.

In 1976, significant funding was directed to a DPM program that provided planning assistance for future proposed APMs in the central business districts (CBDs) of selected cities across the country. The construction of DPMs in Detroit, Jacksonville, and Miami benefited from the DPM demonstration effort and discretionary capital. Today, no sign indicates that the U.S. transit industry is seriously interested in or contemplating with automated transit or PRT applications largely because there is no public policy that encourages the development or deployment of automated transit applications.

The transit industry interest in automated transit is not widely spread in Europe but some scattered countries and/or locations. Besides the DLM applications introduced earlier. Heathrow International Airport in London is ordering six more vehicles for the existing Ultra PRT application and is committed to a second, larger project in the near future (Ultra Global, 2015). Working alone, without significant local and federal public agency involvement, a private investor cannot implement a project in a city or an urban area, even if it looks profitable. A recent survey indicated that there are large engineering procurement and construction companies that are preparing to submit, or already have submitted, unsolicited proposals to build and operate PRT systems. There are also examples of successful collaborations between public and private sectors as documented in the earlier chapters. However, no large-scale or full-fledged automated transit application can be implemented in urban areas without the support of local government, which are often governed or influenced by the national policy.

The FTA, the successor of UMTA, has been absent from funding the planning and construction of automated transit systems after the implementation of the Morgantown GRT and the three DPM applications in Detroit, Jacksonville, and Miami. The only direct involvement with automated transit for FTA has been the National Transit Database (NTD) data collection effort from those applications that have been receiving federal funding. After the projects in the 1970s and 1980s, only a few institutions at the national and/or international level continued their efforts in researching and helping to plan and better understand automated transit development processes. More institutions, professional organizations, and/or trade unions are needed to spread the message, advance the technology, and explore the potential for implementations of automated transit.

Given the complexity of automated transit technology and intensity of capital investment, it is critical to have support at national/international levels for automated transit applications to take place. The support may not necessarily be financial, even though it is certainly welcome. The development and promotion of public policy at the federal level in the United States that help fund an automated transit research/testing program would be most germane. The current political and administrative environment seems fairly conducive to the development of such program. For example, FTA has recently gained authority in safety oversight of transit systems, which is a significant step toward introducing a deeper and more detailed safety culture to the public transit industry in this country. Utilizing its newly acquired safety oversight role, FTA could also choose to explore the safety improvements brought on by automated transit applications.

As suggested by Sproule and Leder (2011), the impetus for the development of automated transit applications in the United States has already been planted in the Reuss-Tydings amendments to the Urban Mass Transportation Act of 1964 (U.S. Congress, 1964). These amendments once required the Secretary of Housing and Urban Development to:

"Undertake a project to study and prepare a program of research, development, and demonstration of new systems of urban transportation that will carry people and goods within the metropolitan area speedily, safely, without polluting the air, and in a manner that will contribute to sound city planning. The program shall concern itself with all aspects of new systems of urban transportation for metropolitan areas of various sizes, including technological, financial, economic, governmental, and social aspects; take into account the most advanced available technologies and materials; and provide national leadership to efforts of states, localities, private industry, universities, and foundations."

The newly authorized Fixing America's Surface Transportation (FAST) (U.S. Congress, 2015) Act may also provide additional tools and flexibilities for FTA and other federal agencies to assist the development of automated transit applications. For example, the clear mandate on "establishing a pilot program for communities to expand transit through the use of public–private partnerships" may encourage the automated transit applications that serve certain transit oriented development (TOD), which may have the potential to generate decent investment returns for private investors, who can in turn fund or contribute to the automated transit applications.

On the other hand, since our society has already spent billions of dollars and built millions of miles of roads and bridges in the past century, which garnered little outside criticism of the expenses but were often proudly claimed as fruition of civilization and/or engineering wonders, it may just be possible to layer PRT guideways on top of existing roadway networks and replace private automobiles with PRT or APT vehicles.

Others who seek more progressive solutions believe that PRT is capable of adapting to existing patterns of living and working, whereas line-haul transit is only efficient in corridor developments. In a large number of metropolitan areas around the world, urban roads are already congested and land availability and cost forbid any road expansions. With a much smaller footprint and a fraction of life-cycle costs of conventional transits such as light rail transit (LRT), subway, and commuter rail, PRT may be able to combine the benefits of both private automobiles and public transit by providing a no-wait, well-connected, origin-to-destination one-seat ride for most urban dwellers.

Looking back, it is clear that the national need of shifting military capacity to civil use has resulted in the overdesign, overbuild, and overpromises of AGT or PRT applications, which may have contributed to larger costs than necessary and eventually contributed to the negative perceptions of AGT applications in the United States. Additionally, lack of solid public policy and continuing research and demo funds has contributed to the lack of fully understanding the benefits of AGT. Public policy may spur rapid and devoted resources but may also damage the result when under tight deadlines and political pressure.

9.3 DESIGN MATTERS

Unlike traditional intercity passenger rail and commuter rail, which are regulated by the Federal Railroad Administration (FRA), and conventional transit, which is overseen by the FTA but not regulated, automated transit applications do not have a clear jurisdiction at this time in terms of federal safety regulation and enforcement. The National Transportation Safety Board (NTSB), without formal authority to regulate and originated from aviation accident investigations, can only investigate and make recommendations after serious accidents. Automated transit applications, especially such systems as AGT, are inherently complex that involve multiple interacting subsystems, new technology, and public safety. It is essential to establish minimum standards for their design, construction, operation, and maintenance.

As mentioned in Chapter 2, Dr. Anderson, a key member of the Economics Evaluation Panel of Automated Guideway Transit in the 1970s (Office of Technology Assessment, 1975), voiced his firm belief in the feasibility of PRT technologies via many PRT conferences and publications. He also noted many engineering companies with a history of defense contracting had a tendency to design much heavier AGT vehicles than necessary, which in turn demanded a guideway twice as wide and as deep as that dictated by load factors. Dr. Anderson's observation was encountered in the Morgantown GRT and repeated in quite a few failed or widely criticized PRT and DPM applications.

The tendency for exaggerated design in the 1970s and 1980s may have stemmed from the lack of understanding of benefits created by small vehicles with smaller footprints. It might also be motivated by acquiring extra "insurance" for newly invented automated transit technologies. The lack of design standard did not help correct those misgivings, which resulted in larger infrastructure costs than necessary and brought bad reputation for all automated transit applications.

Past experiences showed that lack of standards may open the doors for altering the technology specifications, which may eliminate the advantages of certain technologies altogether. Looking back to the various stages of automated transit development, especially PRT applications, many people believed that the PRT concept is feasible technologically, despite the doubts and negative publicities. With more than four decades of safe operations and sufficient safety performance measures, AGT applications have been proven safer and more secure than any other guideway or roadway transit. The unexpected costs have often been associated with exaggerated design, expedited construction, and testing/demonstration nature of early automated transit applications. Concerted industrial design standards are necessary to help minimize costs and take advantage of the latest technological advances.

9.4 DEMONSTRATION PROJECTS ARE NEEDED

Many locations, such as Cincinnati, OH (1996); New Jersey (2007); Ithaca, NY (2007); and San Jose, CA (2009) have evaluated the viability of PRT in various urban, regional, even state wide applications. Most of the studies gathered information on technology suppliers and related literature and applications, some of them estimated ridership, capital or operation and maintenance costs. Few had advanced to stages of design as in Chicago (1991) and none had reached procurement stage.

As explained in the early chapters, all three PRT applications in Heathrow, Masdar City, and Suncheon are essentially shuttles and embody PRT functionality to a rather limited extent. There are still long way to go to reach a network system that will test, prove, or realize the full potential of PRT applications. It seems that the PRT development is currently stuck between a rock and a hard place: there is not enough market interest for a full-fledged system to be implemented while no agency is willing to procure a full PRT system since there is not a proven application.

Will PRT be the next dominating transportation mode of the century? Quite a few dominant "authority figures" were quick to dismiss the PRT idea as "inherently unsound," but the idea resurges every two decades or so, and there are currently more than one and half million entries on the Internet that are directly related to PRT. Are those scattered ideas or initiatives sparkling enough to make us pause, think, and explore? When remembering the "crazy" labelling or humble beginnings of many modern marvels or garage entrepreneurs, such as airplanes and Amazon, you might be able to derive an answer or verdict for automated transit fairly.

As observed in the historical development processes, there are quite a few demonstration project/feasibility studies of AGT under the national development policy in the 1970s. For example, local government agencies in Denver CO; St Paul MN; San Diego and San Jose in CA all participated in the feasibility study and submit bid for the pilot project (Office of Technology assessment, 1975). All three built DPMs have gone through the planning process as part of a special demonstration program in the 1970s (Sproule, 2009). Many corridor studies were conducted by different authorities and private interests, such as Port Authority of Allegheny County, Pittsburgh, PA (TERL) project and El Paso /Juarez international link. Given the short institutional memory, it is imperative to document the studies and methodologies for the continuity of automated transit development and progress.

Building on past feasibility studies and current technology development, a pilot or demonstration project for automated transit will not start from ground zero but will certainly carry the deployment potential to new heights. As mentioned earlier, many transit executives and staff have been repulsed by various simulations and/or artistic renderings of proposed systems. A truly convincing evidence can only come from a real automated transit vehicle that they can ride on, a central control center they can operate and/or a guideway system that they can physically exam and decide how well it co-exist with the surrounding environment.

REFERENCES

Langer, G. 2015. "Poll: traffic in the United States." ABC News. Available at http://abcnews.go.com/Technology/Traffic/story?id=485098. Accessed on February 13, 2015.

Office of Technology Assessment. 1975. "Automated guideway transit: an assessment of PRT and other new systems." Prepared at the Request of the Senate Committee on Appropriations Transportation Subcommittee, NTIS Order No. PB-244854, June 1975.

Sproule, W. 2009. "Somewhere in time—a history of automated people movers." In: Proceedings of Automated People Movers 2009, American Society of Civil Engineers, Atlanta, GA, pp. 413–424, 2009.

Sproule, W. J., and W. H. Leder. 2011. "Downtown people movers—history and future in the U.S. cities." In: Proceedings of APM- ATS Conference by ASCE, Paris, France, May 22–25, 2011.

U.S. Congress. 1964. Public Law 88-365, 1964.

U.S. Congress. 2015. The Fixing America's Surface Transportation Act or FAST Act, 2015.

Ultra Global Inc. 2015. "Ultra provides environmentally sustainable 21st century transport solutions." Available at http://www.ultraglobalprt.com/about-us/. Accessed in March 2015.

CHAPTER 10

FUTURE DIRECTIONS

Learning from past experiences in the United States and best practices in other parts of the world, most transportation professionals are optimistic and ready to march toward the "Promised Land" of automated transit applications as part of the intermodal transportation system. For the few "naysayers" along the way, automated transit application will stay as a mirage until they are personally experienced or brought into the true "destination." This chapter outlines a few future directions, which are critical to expedite development processes or avoid any sidetracks or missteps for automated transit development and implementations.

Practical engineers and rational planners understand that a single mode does not solve all urban transportation problems; every mode has a place in the mobility spectrum. The applicability of any particular design/mode is influenced by a variety of factors, such as changing technology, economic conditions, development and demographic patterns, and social acceptance at particular times. Any entity that is contemplating the idea of automated transit applications, or any other form of emerging technologies, must undertake systematic research and demonstration testing to fully understand the advantages and disadvantages of respective technologies. It is likely that automated transit development will evolve along the following directions.

Automated Transit: Planning, Operations, and Applications, First Edition. Rongfang (Rachel) Liu.
Copyright © 2017 by The Institute of Electrical and Electronic Engineers, Inc. Published 2017 by John Wiley & Sons, Inc.

10.1 GROW AUTOMATED TRANSIT APPLICATIONS

As observed or experienced by many airline passengers, the larger and higher capacity automated people mover (APM) applications have become the normal mode to serve busy and growing airports and "airport cities." Airport city is a relatively recent concept that includes a number of logically combined elements that reinforce each other. The Airport APM (AAPM) applications not only move travelers easily through the airport process but also connect various functions and meet the individual needs of travelers to the extent possible. AAPMs are no longer relegated to the peak-hour ridership of a few thousand passengers but become a routine presence for airport systems that must carry 9000 to 10,000 passengers per hour per direction (PPHPD) during peak hours (Lindsey, 2001). For example, Hartsfield–Jackson Atlanta International Airport, Washington Dulles International Airport, and Dallas-Fort Worth International Airport are all operating along this high capacity range.

Besides the widespread AAPM applications in the United States, more such applications are springing up in many international airports around the world. For example, Beijing Capital International Airport opened its APM system in time for the 2008 Olympic Games. Mexico City International Airport, Charles de Gaulle International Airport in Paris, and Toronto Pearson International Airport all have just opened their APM systems within the past few years.

Additionally, after successful conversion of the Paris Metro Line No. 1 and initiation of Line No. 14 as driverless metros (DLMs), Paris Metro has already started its next fully automated DLM application: Line No. 4 and Line No. 15 (UITP, 2013). Many other international locations, such as Beijing, Singapore, and San Paulo, are also in the process of planning and implementing fully automated DLM.

If the historical development and die-hard ideas of personal rapid transit (PRT) applications have been torn between our desire for sustainable transportation solutions and clinging to the comfort and convenience of private automobiles, the new PRT applications that are capable of reducing congestion and air pollution while providing direct origin to destination services at any time of the day become increasingly real and appealing, as they are impelled by modern communication and control technologies and "Star Trek" quality images. With very short or zero wait time—the PRT vehicle would wait for people rather than requiring passengers wait for vehicles—the quality of service will mollify any stubborn opposition.

It is comforting to see the number of DLM and APM applications grew rapidly in many international locations and the first automated transit application in the United States, Honolulu Rail Transit, is under construction. With the heightened interests in automated vehicles from private sectors and general public, it is possible and necessary for the U.S. government to take the lead in supporting more automated transit applications, among which, automated guideway transit (AGT), such as APM, DLM, and driverless LRT (DLLRT) may offer the fastest track for implementation as they possess mature technologies, tested markets, and experienced system suppliers. Comparing to the automated bus, which are still in the process of testing and perfecting navigation, recognition, and guidance functions, AGT applications expose less safety risk for transit agencies while providing much higher capacity and service quality.

10.2 CREATE NEW MODE

As defined in Chapter 1, a brand new automated transit mode, which may be called "automated personal transit (APT)," might become a reality when autonomous vehicles are enabled by advanced technology and their ownership is transferred to public agencies. Such APT systems will possess all the advantages of private vehicles, which will take travelers from door to door without any waiting, transferring, or sacrificing privacy during their journey from their origin to their chosen destinations. The cost of owning, maintaining, and storing a private vehicle, which is used only a couple of hours each day, has the potential to push a large portion of households toward APT. The wide availability and lower cost associated with APT have the promise to attract a majority of households toward the shared economy and shared mobility by using APT. Riders can rely on their cell phone or computer systems to hail an APT vehicle when it is needed. There will still be a small number of individuals who choose to own and exclusively operate their own vehicles, but the majority of the traveling public could choose to give up automobile ownership, not only because of high cost of owning but also because of the convenience of getting a vehicle when it is only needed even it is not owned by the traveler.

The objective of this newly formed automated transit application, APT, is to overcome the shortcomings of conventional transport modes, meet future requirements of next generation travelers, and enable the prosperity of shared economy and shared mobility. It might be difficult to describe such a new system advent its arrival, especially when the expectations for such a new system vary greatly. However, it is possible to identify three basic objective factors for the new system: the need to satisfy a broader range of technical specifications; the need to avoid negative social effects, such as traffic accidents, environmental hazards and pollutions; and the need to operate the system as part of a comprehensive and fully integrated public transportation and communications network.

It may seem confusing first that both "personal" and "transit" are used together to describe a transportation mode, as most readers connect "personal" with "private" or "individual"; and "transit" with "public," "large group," or "inflexible." In this particular mode, these two words capture and highlight precisely the beauty and advantages of both automated vehicles and PRT, and therefore APT.

Transit, based on an early definition, comprises all transports in which riders are free of both driving responsibility and social obligation to the driver if any, who is hired for this work (Fichter, 1964). Similarly, an early definition of private transport modes explained that a vehicle's prime user retains the custodial relationship as well as the responsibility of driving, with its implied demand for skill and constant attentiveness. The trip purposes are often personal or incidental to employment. Even riders in the private vehicle have a social or business relationship to the driver (Fichter, 1964). Examples include parents dropping off a child at school, or a few friends carpooling together to a movie.

When a driver is no longer needed for automated or driverless cars or transit vehicles, the passengers or users are liberated from the relationship or obligation that defines private transportation. When that happens, it is not difficult to imagine that APT mode will stand for automated cars, owned and/or operated by public agencies or their subcontractor entities, to serve individuals when and where they are needed. By then, the "personal" phrase depicts "personalized," "individual" services; and "transit" means the users are free from the obligations of owing, operating, or maintaining the vehicles and associated systems.

Prominent corporations both within the traditional auto industry and those outside are developing an increasing number of intelligent autonomous vehicles. "They may evolve as consumer vehicles those individuals own. Or perhaps not!" hypothesizes Brownell and Kornhauser (2014), Princeton University. Why not just subscribe to a mobility service that you can summon? Several major international corporations expect to have commercially ready consumer vehicle products for on-street operation within a few years. Early deployments of 100 vehicles are planned at the University of Michigan, Ann Arbor as part of a decade-long program, and the University of West Florida, Pensacola has planned a system within a much shorter timeframe. Similar programs were announced by the UK's Automotive Council for Milton Keynes and by Volvo for Gothenburg, Sweden, in late 2013.

Speculating the future of automated vehicles, most planners and analysts believe that a portion of the population or transportation users will disown the private vehicle as the modern communication and automated transportation technology develops to the extent that a user may easily summon a vehicle with minimum waiting, and the need for owning a private vehicle diminishes or the cost reductions outweigh the convenience for owing, maintaining, and finding space to store the vehicle. By then, PRT and AV will morph into a brand new hybrid mode: APT. PRT may still have its niche to operate podcars along dedicated guideways in congested urban areas while APT may have the potential to replace the majority of private conventional vehicles. In that future era, the taxi could be replaced by APT all together, which will forge a different operating environment for a Taxi.

10.3 CONDUCT FURTHER RESEARCH

Serving as critical links in many large airports, dense downtown areas, and major activity centers (MAC), AGT applications around the world have been performing the vital function of connecting passengers to and from their origins to their destinations every day. However, since most of the AGT applications are short in length, ranging from a few hundred feet to a few miles, generally confined to the environs of airports, and/or owned by private operators, their importance or vitality is often ignored or taken for granted. The lack of significant research from academic institutions and absence of testing and demonstration programs for automated transit applications have severely inhibited its development.

In recent years, the only two systematic analyses of APM applications are funded by the Airport Cooperative Research Program (ACRP) and focused solely on AAPM applications (Gambla and Liu, 2012 and Lea + Elliott, 2010). There is no substantial research or verified performance measures that outline accurate benefits and costs of AGT applications in urban environment, not to mention their impact on surrounding communities. Lack of research, testing, and demonstration has inhibited the implementation of AGT in urban environments. Even though AGT system designers may have good cost estimate for AAPM applications, which may not be directly applicable to urban environment, the benefit data are almost non-existent.

A quick scan of the existing literature and ongoing automated transit studies reveals that the specifications of technology and assessment of costs may be relatively straightforward, but quantifying benefits associated with the implementation of a transportation project and evaluating the market conditions are complex. There are a number of analytical tools to assign a dollar value to benefit, however; some impacts, such as congestion relief, safety improvement, and air quality improvement, are often difficult to quantify or monetize. Other traits or characteristics, such as aesthetic appearance, may not even be quantifiable. Environmental and societal impacts are often referred as *external* effects of transportation activities because they are not reflected directly in monetary costs and benefits of project implementation. By externalizing these factors, cost-benefit analyses often do not capture the full value of beneficial impacts, even those external factors may be significant, sometimes, even fatal to the implementation of automated transit applications.

On the other hand, the most difficult task so far to convince decision makers or the public of the feasibility of automated transit applications is often the estimation of ridership in the absence of revealed preference (RP) data. While many studies may use stated preference (SP) data before a real-world application may take place, the biases associated with SP data are well known, and the discrepancies between the two are difficult to estimate. Further complicating automated transit ridership estimates are the intermodal transfer penalties associated with initial applications. When starting from a short segment, an individual corridor or some other type of limited-scale alignments, which are no different from the initial segments of any other transportation projects, automated transit ridership estimates may suffer from lower values owing to its limited network coverage and heavy intermodal penalties.

The premise for the design of most automated transit systems, especially PRT applications, is the direct origin to destination travel in an interconnected guideway network, which requires dynamic transit routing and dispatching. If vehicles/trains were dynamically coupled and decoupled as they progressed through a transit way system, and if off-line stations were created to facilitate demand-responsive service, the operational costs can be reduced, travel time decreased, and level-of-service improved significantly. The task to dispatch a large number of automated podcars or PRT vehicles among a dense network of automated transit application remains a great challenge, which may also be the "catch 22" for the doomed expansion of PRT systems. There is no large-scale PRT application in existence, which made it difficult to develop or test the dynamic dispatching of large number of small PRT vehicles among dense transit networks. Meantime, many entities are shying away from the risk or uncertainties associated with potential issues created by the dynamic dispatching. In the author's opinion, this may be one of the limited areas where more and further simulations are needed but simulation should not replace or sacrifcie the real testing facilities.

Furthermore, it has been advocated that the transit industry provides the best environment to test automated roadway vehicle operations. The larger the network, the more complex trip scheduling and empty vehicle management become. In transit operations, the complexity increases with the square of the number of stations. In roadway vehicle operations, the number of "stops," equivalent to "transit stations" can become finite and/or even dynamic, and sometime, illogic. One immediate question is "will surges in demand overwhelm real-time scheduling and fleet management functions of the control software?" Or "how will the system respond to perturbations—such as accidents, power outages, fallen trees, medical emergencies, criminal and terrorist acts, etc.?" The software programming and communication and control applications may need to be designed and implemented for large network and geography-specific automated transit configurations. As concluded by Aerospace Corporation (Page, 2012), significant work is required for the development, validation, and verification necessary for large meshes of automated transit, especially PRT applications. Federal funds would be well justified by the foreseeable benefits. Given the high priority placed on reliable, safe, and secure service options, the need for investment in advanced software development and communication control systems may come from either public or private sector.

10.4 SPONSOR DEMONSTRATION PROJECTS

As part of the assessment of AGT technologies, the Office of Technology Assessment (1975) proposed demonstration projects in the 1970s. Anticipating a fleet of automated vehicles owned by public agencies or private entities, the exact same model as APT even the term was not used, the same office then proposed rental arrangements of special small vehicles to obtain better utilization and to minimize storage problems. A variety of options could be available, but essentially the vehicles would be rented by individuals from a private company or public agency for single trips or extended periods of time. Utilizing conventional manually operated vehicles, the concept is identical to Zipcars or other similar transportation network company (TNC) applications.

Even in the 1970s, such an arrangement was in operation in Amsterdam (Office of Technology Assessment, 1975), where one could rent at 4¢ per minute, small battery-powered vehicles, not unlike golf carts, for transportation to various places within the city. Special parking places were set aside for these vehicles at recharging stations near major attractions. By the end of 1975, there were plans to have 15 stations and 125 cars in service. A similar operation has been visualized as a demonstration project in Washington, DC, for transportation between the many tourist attractions along the National Mall and elsewhere in the heart of the city. Remote parking for full-sized family cars could be provided at locations such as Robert F. Kennedy (RFK) Memorial Stadium and the Pentagon on weekends. Small vehicle rental and storage facilities available at these locations, selected metro stations, and the Visitor's Center at Union Station could provide a personal transportation service. Of course, such demonstration project did not come to fruition when the national priority shifted away from automated transit applications.

In contrast to the full-scale applications of DLM and ubiquitous APM systems around the world, there is still no full-scale and/or fully functional PRT application in the world as of 2016. As mentioned earlier, the few pioneer

PRT applications in London Heathrow International Airport, Suncheon Bay, and Masdar City are all simple shuttles or line-haul operations that have neither network scale nor direct origin and destination operation. As we all understand that the basic requirement for safety and reliability standards can only be tested and validated when a real application is put into place. Not many entities or transit agencies are willing to make the commitment before solid or comfortable cost ranges and related ridership estimates are provided, which again will be possible only with adequate testing, demonstrations, and real-world applications. It seems that the time is ripe for a leader, be it government agency or private entity, to emerge and lead the way for automated transit development by enacting and supporting demonstration projects.

10.5 DEVELOP PERFORMANCE MEASURES

The recent development in standardization for automated guideway transit operation by various professional organizations may serve as evidence of the APM/ATS industry's maturity (Lott, 2014). The International Electro Technical Commission (IEC) has adopted the Automated Urban Guided Transport Safety Requirements (IEC 62267) while the American Society of Civil Engineers (ASCE) has developed the Automated People Mover Standards (ASCE-21). The ASCE APM Standard has recently added off-line station provisions, which is quite relevant to the design of merging automated roadway vehicle technology into guideway transit through an evolutionary process.

As mentioned earlier, most of the APM systems are short in length, generally confined to the environs of airports, and owned by private operators of the various facilities. There is very limited research, testing, and demonstration of APM systems and almost blank when it comes to the performance measures of such systems. As demonstrated in Chapters 5 and 6, the linkage between performance measures and demand are non-existent. The reason for the lack of connections between performance measures and demand of AGT applications is manifold.

One of the key causes may be that the performance of APM applications, especially AAPM applications, is narrowly defined and measured for existing systems. In many cases, performance measures of AAPM systems are specified in contract documents as a means of verifying the operator's compliance with contractual requirements. Specified performance measures for AAPM typically are based on scheduled and actual operating data, which are reflected by service mode, fleet, station platform door, and system service availability. Widely used in AAPMs worldwide, these availability measures provide the most appropriate level of accountability for the system and its elements that most directly affect the quality and level of service experienced by passengers. However, the true success of transit operations is simultaneously determined by efficiency, balance between supply and demand, and customer satisfaction, which should be incorporated into future performance measures.

Given the shortcomings of the existing performance measures of APM systems, ACRP Project 03-07 has studied and proposed comprehensive performance measuring system for AAPM applications. After surveying the AAPM operators in North America and participating in extensive dialog with various stakeholders, the research team and subsequent studies have developed new sets of performance measures to reflect a comprehensive evaluation of APM services in airports (Gambla and Liu, 2012 and Liu and Huang, 2010). Similar studies or explorations should be executed for various automated transit applications in urban areas.

10.6 ENCOURAGE DIVERSE BUSINESS MODELS

As observed in the case studies presented in the earlier chapters, implementing an automated transit system is a complex public works and technology project with costs in the tens of millions of dollars. Large urban projects, such as ALRT or DLM, may run several years and involve several stages. The larger and more complex the project, the greater the project team, the larger the number of stake holders, and the stronger the technical and organizational skill requirements. In a nutshell, the implementation process for an automated transit application may involve several or all of the stages listed below:

- Planning;
- Design;
- Engineering;
- Permitting, including environmental impact assessment;
- Site preparation, including utility relocation;
- Construction, including impact mitigation, landscaping, civic embellishments and artwork;
- System installation;
- System integration and safety certification;
- Training and launch of revenue service;
- Operations and maintenance.

Presently, there is no clear business model for automated transit implementations. As demonstrated in Chapter 8, various business model structures for different submodes of automated transit technologies have been observed in the real world and their success and failure varied widely. If automated transit applications are designed to feed and reinforce existing transit, such as recently converted DLMs in Paris, the funding or operation support may originate from the general transit operation and/or subsidy channel. As an organic portion of a comprehensive transit system in an area, automated transit applications may be selected or converted as an improvement to the conventional transit technology rather than a deviation. The downside of such business model is that financial or accounting systems often mix revenues and expenses from both conventional and automated transit together, so it may be difficult to distinguish or recognize how cost-effective the automated transit applications are.

Experiences from privately funded APM projects also mixed both positive and negative in a different way. For example, Huntsville campus in Alabama, University of Indiana hospital complexes, and the Getty Museum in Los Angeles are often cited as positive examples while Harbor Island in Tampa, FL; Wellington in Boston, MA, Las Colinas in Texas, and Oeiras in Lisbon, Portuguese are negative examples for various reasons. Similar to the reputation of public owned and operated applications, the private implementation of APM systems could not escape from the criticism of high cost, low ridership, and/or many unmet expectations, either. In the context of private owner and private operator business models, many questions have to be answered before investors open their wallet.

- Can the flexibility of automated transit configuration bring more transit accessibility or a large rise in property values that helps pay for the cost?
- How substantial will be the revenues from advertising, recharging, and in-guideway utility conduits?
- What can be learned from the growth of car rental and ride-sharing communities?

As suggested by many researchers (Fagnant, 2015 and Brownell and Kornhauser, 2014), the emergence of automated vehicles holds great promise for the future of transportation. While commercial sales of fully self-driving vehicles will probably not commence for several more years, which leaves room for PRT to prosper if workable business models are developed. As described in Chapter 8, private entities or public-private partnership (PPP) arrangements may hold more promising future in funding and supporting PRT applications, if such configurations can be designed to provide satisfactory answers to the questions raised earlier.

Will automated transit be the next dominating transportation mode of the century? With an open mind and out-of-the-box vision, some transportation professionals believe that for PRT to become a reality, it may require a revolution in the way we live and travel. That is, PRT may not be feasible if highways and private automobiles continue to be our anchor mode of transportation in the near future.

Further down the road, once automated vehicle becomes a reality, a new transportation mode, APT, that combines the advantages of both private vehicles and public ownership for personal travel will arrive. This new mode is the shared autonomous, or fully automated, vehicle (SAV), combining features of short-term on-demand rentals with self-driving capabilities: in essence, a driverless taxi or APT as defined in earlier chapters.

It is quite conceivable that a fleet of automated vehicles could be owned by an entity, such as Uber, or a public transit agency such as New Jersey Transit. Travelers may simply purchase a ride when needed without any burden of maintaining or owning the vehicle. A shared fleet of automated vehicles, or automated taxis, will eliminate significant driver costs to make the taxis much more affordable (Brownell and Kornhauser, 2014). Re-examining the individual modes of automated transportation presented in Figure 1.3 in Chapter 1, PRT may merge with automated vehicle to form APT and provides transit services under a different business model.

When many urban economists, entrepreneurs, and public policy analysts work together to examine these issues thoroughly, satisfactory answers or solutions can be found. Equipped with better cost and benefit information, investors have a good chance to make educated guesses or informed decisions so their expectations can also be brought in-line with results that can be realistically accomplished by the proposed automated transit applications. When many problems and pitfalls encountered before can be avoided or mitigated in the new round of implementations, the benefits can be enormous.

10.7 GATHER PUBLIC SUPPORT

As observed by Jain et al. (2014), there is a rapid increase in the use of personal modes of transport over public transit all over the world. In spite of tremendous efforts to promote public transport by authorities, they still fail to attract the attention of a great proportion of the masses due to various reasons. PRT is an efficient rapid transit system that can provide the last mile connectivity to users with a high level of reliability and comfort. It is a demand-responsive system that ensures uninterrupted, point to point journey, between origin and destination, which is missing in conventional public transport systems. However, it failed to garner public support or recognition as a viable transportation mode.

It is important to educate the general public about the true characteristics of automated transit, which is one of the objectives of this book. As pointed out earlier, the lack of public attention or recognition for automated transit applications may stem from its inherited small-scale, non-intrusive operations, and no fatality safety record. However, the very same traits, coupled with small number of applications, also made automated transit applications not available or accessible to majority of travelers. For example, the AAPM can only be utilized by airline passengers and/or airport workers. Certain dedicated air-side, land-side, or transferring shuttles of AAPM application may not even be used by all airport passengers either. Therefore, it is important to bring automated transit experiences to potential users via social media, formal education, and general public outreach.

Another challenge to automated transit applications may also originate from its small scale but can be remediated in due time. As confirmed by previous research (Liu et al., 1997; Office of Technology Assessment, 1975; and Guo, 2007), transit use is often inhibited by intermodal or intro-model transfers, which are the products of intermodal or intro modal connectivity. The concept of a transfer penalty implies that there are two transfer components affecting the disutility of a transit trip. The first component is transfer time, such as waiting time or walking time associated with transfer. The second is simply the requirement to change vehicles, which produce a penalty independent of the time it takes to transfer between different modes or vehicles (Liu, 1996). Transfers are undesirable, especially when many conventional bus and rail service routes frequently involve long waits, prolonged walking distances, and/or at non-weather protected locations.

One solution to the transfer issue is to develop a transit system that carries a passenger from origin to destination on a single vehicle, preferably on a personal basis without intermediate stops, which is precisely what PRT is capable of offering when expansive PRT network is developed. Since it eliminates the need for transfer, PRT is a public transport mode that provides door to door service while minimizing access/egress walking distances. These features of PRT may also make it a state-of-the-art public transport system which has the potential to attract non-captive riders from personal modes to public transport.

A full-fledged PRT network or automated transit network (ATN) may well exceed public expectations but very expensive under current technical capabilities, which may also be the reason that only limited scale or functions of PRT applications are put into places in recent years. As the first step, PRT actually reduces the intermodal and/or intro-model penalty by reducing the wait time. For example, the Ultra podcars in Heathrow International Airport are usually at the station waiting for passengers most time, which basically reduce the waiting time to zero and reduce the transfer penalty significantly. In certain MAC locations, PRT or APM is implemented to reduce walking time or eliminating walking all together, which reduces the transfer penalty also.

It is commonly accepted that introduction of PRT in an area leads to the improvement of the accessibility of the area. However, studies in Denver indicate that the negative environmental impact of guideway interchanges significantly outweighs the marginal patronage advantage achieved by eliminating transfers. Investigation may show that what is really important to those currently depending on autos and pressed by increasing auto costs is safe, reliable, frequent service to multiple destinations in a reasonable period of time. Transfers that can be accomplished quickly in a climate controlled, secure location may be quite acceptable. If transfer points co-exist with opportunities to turn single-purpose trips into multi-purpose trips, such as pick up dry cleaning, drop off/pick up children or run errant, so much the better.

According to an earlier study (Young and Muller, 2012), PRT systems function best in a uniform gridded network. In such systems, particularly those using one-way guideways, passengers will be afforded a uniformly high level of accessibility to all sections of the urban area. Many PRT advocates feel that ubiquitous service and coverage is the most important attribute of the PRT concept; so that only by providing service to the majority of dispersed trip ends in the urban area can PRT be an effective transit competitor to the auto. The limited-scale PRT applications seen today could not represent the true function of PRT but the first and necessary step for PRT network development.

In many respects, PRT and auto networks share similar attributes. Each involves small vehicles with low vehicle occupancy. Each functions best on a guideway or roadway system with relatively even spacing. Neither favors online stations, and both seem to have low tolerances for congestion. The similarity between PRT and the auto is a conscious effort to emulate the most "successful" transport mode history has ever seen. Yet how far should this similarity go? Some argue that by replacing autos with a transit system so similar in operating characteristics, we run the risk of propagating many of the adverse impacts of the system we seek to distance, including the low load factor typical of autos. The parallel will extend to the future when automated vehicles and APT mode become reality as DLM and APM still represent the high capacity, high density public transit services while APT the low capacity, flexible individualized alternatives.

The traction on vehicle automation can be proven by both formal events and casual observations. There are about 51 million entries on the Internet under the search for "vehicle automation" and 40 million for "automated transit." The automated transit and shared mobility track for the annual Vehicle Automation Symposium (http://www.automatedvehiclessymposium.

org/home) have been attracting large number of audiences, since its first year in 2011. The Transportation Research Board's Automated Transit Systems (ATS) Committee (AP040) has received more and more research papers and presentations from year to year on the topic of automated transit spanning wide ranges of issues, including those involving legal, ethical, planning, operations, and social impacts.

As anticipated by the World Economic Forum (2015), a non-profit international organization, vehicle automation will have the potential to save more than 30 thousand lives in the United States alone by avoiding accidents created by human errors. When the driving task is completely delegated to the machine itself, 75 billion hours of commuting time can be reclaimed or directed to other productive or pleasant use. As part of the effort to fully realize such potentials, the automated transit community has been working with multiple stakeholders in developing a strategic roadmap.

Working toward a common goal, many sectors, such as automotive, insurance, technology and government, can and must play critical roles in the automated transit development process. Technology firms must adapt to long product cycles and high safety regulations. The Original Equipment Manufacturer (OEM) and supplier companies must assist and prepare for technology players and prepare for liability residing with someone other than the human driver. Regulatory agencies must address the development of relating software, as opposed to just hardware. Government agencies, such as Federal Department of Transportation (USDOT), National Highway Transportation Safety Administration (NHTSA), and state Department of Motor Vehicles (DMV), need to act swiftly to clarify the safety hurdles. Insurance companies may need to accept potentially lower auto premiums over time and provide liability coverage to someone other than the human driver.

As the ultimate beneficiary of automated transit systems deployment, travelers will truly enjoy the mobility and freedom brought on by automated transit applications. Cities and suburban and rural areas in the future may have and enjoy various automated transit deployment models and/or schedules. Such automated transit deployment across the country has the potential to provide better mobility and accessibility for all, no matter where the person lives or his /her income and/or driving ability.

REFERENCES

Brownell, C., and A. Kornhauser. 2014. "A driverless alternative: fleet size and cost requirements for a statewide autonomous taxi network in New Jersey." *Transportation Research Record: Journal of the Transportation Research Board*, vol. 2416.

Fagnant, D. 2015. "Operations of a shared autonomous vehicle fleet for the Austin, Texas, Market." In: Proceedings of 94th Transportation Research Board Annual Meeting, Washington, DC, 2015.

Fichter, D. 1964. *Individualized Automatic Transit and the City.* Providence, Rhode Island.

Gambla, C., and R. Liu. 2012. "Guidebook for measuring performance of automated people mover systems at airports." Airport Cooperative Research Program (ACRP) Report 37 A. Sponsored by the Federal Aviation Administration (FAA). Transportation Research Board (TRB) of National Academies. Washington, DC.

Guo, Z., and N. Wilson. 2007. "Modeling the effects of transit system transfers on travel behavior: case of commuter rail and subway in Downtown Boston, Massachusetts." *Transportation Research Record: Journal of the Transportation Research Board,* vol. 2006, pp. 11–20.

Jain, U., P. Sarkar, and A. Vibhuti. 2014. "Impact of personal rapid transit on accessibility index." TRB 93rd Annual Meeting Compendium of Papers. Washington, DC.

Lea + Elliott. 2010. "Guidebook for planning and implementing automated people mover systems at airports." ACRP Report 37, Transportation Research Board, Washington, DC.

Lindsey, H., and D. Little. 2001. "Driverless rapid transit systems take hold." In: American Public Transportation Association (APTA) Rail Transit Conference, Miami, FL. June 2001.

Liu, R. *Assessing Intermodal Transfer Disutilities.* University of South Florida, 1996.

Liu, R., and Z. Huang. 2010. "System efficiency: improving performance measures of automated people mover systems at airports." (10-0835). *Proceedings of Transportation Research Board (TRB) 89th Annual Meeting,* January 2010, Washington DC. Peer reviewed.

Liu, R., R. Pendyala, and S. Polzin. 1997. "An assessment of intermodal transfer penalties using stated preference data." *Transportation Research Record: The Journal of Transportation Research Board,* vol. 1607, pp. 74–80. Peer reviewed.

Lott, S. 2014. "Implications for guideway/transit way and station design with respect to automated transit vehicles" In: The 3rd Road Vehicle Automation Conference, San Francisco, 2014.

Office of Technology Assessment. 1975. "Automated guideway transit: an assessment of PRT and other new systems." Prepared at the Request of the Senate Committee on Appropriations Transportation Subcommittee, June 1975. NTIS Order No. PB-244854.

Page, P. 2012. "Look, no hand." *Container Management.* Baltic Publishing Ltd, pp. 36–38.

UITP. 2013. "A new conversion: Paris Metro will automate Line 4." Available at http://metroautomation.org/a-new-conversion-paris-metro-will-automate-line-4/. Accessed on July 15, 2013.

World Economic Forum, 2015. "The outlook of the United States." Available at https://www.weforum.org/events/world-economic-forum-annual-meeting-2015/sessions/outlook-united-states. Accessed in August 2015.

Young, S., and P. Muller. 2012. "Exploratory concepts in advanced transport: mobility and land use enhancements enabled by automated small-vehicle transport technologies at the village west development—a personal rapid transit case study." Monograph.

INDEX

Automated Transit: Planning, Operations, and Applications, First Edition. Rongfang (Rachel) Liu.
Copyright © 2017 by The Institute of Electrical and Electronic Engineers, Inc. Published 2017 by John Wiley & Sons, Inc.

IEEE PRESS SERIES ON SYSTEMS SCIENCE AND ENGINEERING

Editor:
MengChu Zhou, *New Jersey Institute of Technology and Tongji University*

Co-Editors:
Han-Xiong Li, *City University of Hong-Kong*
Margot Weijnen, *Delft University of Technology*

The focus of this series is to introduce the advances in theory and applications of systems science and engineering to industrial practitioners, researchers, and students. This series seeks to foster system-of-systems multidisciplinary theory and tools to satisfy the needs of the industrial and academic areas to model, analyze, design, optimize and operate increasingly complex man-made systems ranging from control systems, computer systems, discrete event systems, information systems, networked systems, production systems, robotic systems, service systems, and transportation systems to Internet, sensor networks, smart grid, social network, sustainable infrastructure, and systems biology.